LANCHESTER LIBRARY
3 80
D1187782
WITHDRAWN

Make It!

Engineering the Manufacturing Solution

For my two wonderful daughters

Joanna and Claire

£25.00
8/99

Make It!

Engineering the Manufacturing Solution

John Garside

Warwick International Manufacturing Group

OXFORD AUCKLAND BOSTON JOHANNESBURG MELBOURNE NEW DELHI

Butterworth-Heinemann
Linacre House, Jordan Hill, Oxford OX2 8DP
225 Wildwood Avenue, Woburn, MA 01801-2041
A division of Reed Educational and Professional Publishing Ltd

A member of the Reed Elsevier plc group

First published 1999

British Library Cataloguing in Publication Data
Garside, John
Make it!: engineering the manufacturing solution
1. Manufacturing processes 2. Industrial management
I. Title
670.4′2

ISBN 0 7506 4569 5

Library of Congress Cataloguing in Publication Data
Garside, John, 1944–
Make it!: engineering the manufacturing solution/John Garside.
p. cm.
ISBN 0 7506 4569 5 (hbk.)
1. Production planning. 2. Manufacturing cells. I. Title.
TS176.G365 99–30644
658.5–dc21 CIP

Typeset in India at Replika Press Pvt Ltd, Delhi 110 040
Printed and bound in Great Britain by
Biddles Ltd, Guildford and King's Lynn

Contents

[illegible] [illegible] [illegible]

[illegible] L. Ltd Butch
Bundel

Company background.

identification of problems

[illegible]

Introduction

My first book, *Plan to Win,* defined the business processes needed to manage a successful company. However, I realized that the chapter on supply-chain management enabled readers to understand good practice and identify operational deficiencies, but did not provide the tools needed to engineer effective modular manufacturing solutions. So, I decided to write a book, providing a framework on how to do it. It is written for managers charged with the responsibility for improving business profitability and for engineers faced with the challenge of how to create cost-effective manufacturing processes. I have defined the supply chain as spanning deliveries from the supply base, through to in-house manufacturing operations and shipping products to customers, but my main objective is to describe the tools and techniques needed for designing in-house manufacturing systems and supply-chain interfaces to compete effectively in world markets.

Many businesses have made progress towards implementing lean manufacturing through continual improvement programmes and operator participation. Both are important elements in improving performance. However, if a manufacturing system has not been systematically designed by a project team responsible for determining *what* are the business's core operational processes and specifying *how* they will function, it will never realize its full potential. *If the manufacturing system is flawed, without a structured design, it is condemned to remain flawed.* The Japanese manufacturing machine evolved through making licensed products from the West less expensively and of consistently higher quality, by applying its resources to systematic process design. This need to invest resources in designing effective, modular manufacturing systems using dedicated teams remains an issue; many businesses cannot identify the people responsible for establishing how their factory operates. Dedicating teams to design in-house processes provides a real opportunity to introduce a further step change in business performance through implementing engineered manufacturing solutions.

The following logical stages, sequence of events and analyses, essential for designing manufacturing processes and facilities, are covered in my book.

Identify the need for change
Translate the business plan into operational targets that must be achieved through designing core operational processes and confirming that the financial commitments are achievable.

⟶ *Module/cell identification*
Develop a manufacturing strategy and identify the gaps in current operational performance. Establish through strategic make versus buy analysis those activities needed in-house and the module structure considered necessary to implement the manufacturing policy.

⟶ *Steady state design*
Determine the size of modules/cells needed to satisfy market demand. Design how they will operate under steady state conditions, including quality systems, benefits of applying Japanese techniques, introduction of new technology, job requirements and module support services.

⟶ *Dynamic design*
Modify module/cell designs making them tolerant of the dynamic factors impacting operational performance. Select supply-chain and manufacturing planning and control systems, using techniques to optimize performance.

⟶ *Financial justification*
Create a supplier integration policy. Develop and present to senior management the financial case for implementation with detailed costs and time-based plans.

⟶ *Implementation/continual improvement*
Install manufacturing modules/cells and commission equipment. Define the job requirements and select team members, identifying essential training. Introduce continual improvement teams to enhance performance, monitoring achievements through meaningful measures of performance.

The need to undertake a supply-chain and manufacturing design project may be triggered by a number of events. Each may have a different objective, for example reduce operating costs, introduce new products, relocate factories, rationalize facilities, or accommodate significant increases or decreases in production volumes – but all are ultimately aimed at improving business performance and financial viability. Dedicated manufacturing design teams must be established and made responsible for a range of activities; supply-chain and manufacturing systems design, new product and process introduction (including design for manufacture), manufacturing technology development, continual improvement programmes and possibly for instigating similar activities in supplier organizations.

World class supply-chain and manufacturing systems are achieved over several years' development through a combination of identifying business requirements, systematic manufacturing system design, rigorous team selection and training followed by the introduction of continuous improvement teams committed to removing all forms of waste from the operation. *Neither designing manufacturing systems nor continuous improvement programmes alone will ever engineer an optimum manufacturing solution.* These inter-related activities must also be relentlessly driven; introducing step changes through design and refining by continual improvement.

This book is based upon my personal experience of working in different companies, developing effective factories to compete in global markets. However, a significant aspect has been the knowledge gained from the skilful managers and colleagues with whom I have worked. Before being appointed Principal Fellow at Warwick International Manufacturing Group, University of Warwick, in 1997, I had worked for Lucas-Aerospace, Automotive and Fluid Power Divisions, GKN Technology and Dunlop Technology. I was responsible for managing multidisciplinary teams designing supply-chain and manufacturing systems, gaining first-hand experience of precision aerospace and automotive component manufacture, industrial distribution, metal forming and polymer processing. Assignments were worldwide, including the United States and Europe. I hope, therefore, this book will provide a practical guide for people involved in designing manufacturing operations, acting as route map, identifying inter-relationships plus many other essential factors that must be considered when striving to create world class factories.

Acknowledgements

Throughout my working life I have been associated with many talented people in a variety of industries. I have acquired valuable knowledge from managers and colleagues, who have significantly influenced my attitude to manufacturing, shaping the concepts and techniques needed to implement successful supply-chain and manufacturing systems as described here

I wish to thank all those people, and in particular: Professor John Parnaby for helping me gain an in-depth understanding of supply-chain and manufacturing systems, including the methodologies needed to design and implement world class manufacturing systems using a task force approach. Frank Turner for guidance on preparing manufacturing strategies and identifying best practice. Tony Walton for highlighting the importance of people, selection methods, and the impact of team working. Tony Helliwell for giving me insight into the importance of strategic make versus buy policies and creating supplies modules. Tim Bridgman for sharing manufacturing training material encapsulating the experiences we gained in changing factories.

I express my sincere thanks to Professor Kumar Bhattacharyya and the Warwick International Manufacturing Group for providing me with the opportunity and facilities needed to write this book, and also for the chance to disseminate my manufacturing knowledge through working with companies and students associated with Warwick University.

Also my dear wife Ros who has enhanced the overall quality while providing continual encouragement and support.

John Garside

Chapter 1

Manufacturing effectiveness – the need for change

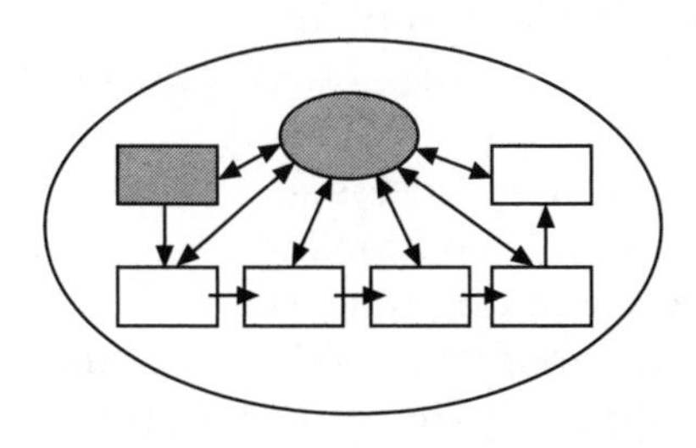

Introduction

Effective manufacturing systems design is the foundation of a competitive and profitable manufacturing business. The supply-chain operation is fundamental to the prosperity of the company, with improvements in operational performance leading to additional profit thus increasing stakeholder confidence and further funds to invest in developing the business. It has been demonstrated by many of the world's leading benchmark companies that designing manufacturing systems to produce high quality products at low cost is the most important factor in delighting customers and growing market share. Technically advanced, desirable products retain a niche market until a manufacturing route is established, making them affordable and more readily available. Manufacturing businesses must invest time and resources into designing how supply-chain and associated in-house manufacturing systems should operate. This design process is generally straightforward, but it does require considerable effort to understand current manufacturing methods, develop innovative processes, capture existing proprietary knowledge and encapsulate them into a robust set of core manufacturing processes. However, companies invariably devote insufficient resources to this most crucial task; often businesses cannot identify the person directly responsible for designing the manufacturing system or being accountable for how the supply chain operates.

In my experience the rigor of this design process underpins supply-chain effectiveness and ultimately business profitability, because of the wide range of critical factors influenced by decisions taken as part of the manufacturing system design process which:

- Identifies appropriate cost-effective production methods.
- Designs manufacturing cells to meet the needs of the business.
- Creates the team environment needed for self-directed work groups.
- Determines the investment required in plant and equipment.
- Decides on the number of people needed in operations.
- Creates the flexibility needed to accommodate changes in production volumes.
- Determines the actual cost of manufacturing processes.
- Introduces systems for controlling materials and tooling.
- Determines the skills required for new working practices.
- Identifies the support activities needed to support the module structure.
- Captures core competencies, embedding them into operational processes.
- Establishes the quality of parts being manufactured.
- Determines the level of customer satisfaction.
- Integrates process design with the product introduction process.
- Improves product integrity with design for manufacture and assembly.
- Assists the supplier base to meet the needs of the business.
- Provides the mechanism for implementing corrective actions and recommendations for continuous improvement.

Redesigning the manufacturing process is not a task that can be allocated solely to supply-chain operational management because its main responsibility is to meet customer delivery commitments and manage resources. Operational managers obviously must have direct involvement and provide many of the resources needed to support the process. The day-to-day responsibility for driving the changes must be given to a project manager accountable for delivering an optimum manufacturing system within agreed budgets and time scales to meet expected customer service requirements and future market demand. However, in my opinion, *a core process team* should be established within the organization to continually design and develop supply-chain and in-house manufacturing processes, supporting both product introduction and operational management groups.

The need to design supply-chain and manufacturing systems may be triggered by several events:

- Reduction of operational costs to match competition.
- Introduction of new product lines.
- Relocation of factories into more appropriate facilities.
- Rationalization of facilities to remove fixed costs and overheads.
- Launch of additional product lines into existing factories.
- Significant increases or decreases in production volumes.
- Improve effectiveness of existing facilities following:
  - strategic make versus buy decisions; and
  - implementation of dedicated core product cells.
- Installation of new production technology for core processes.
- Introduction of new business planning and control systems.
- Significant changes in product technology.

If businesses simply foist the responsibility for introducing these changes onto operational supply-chain management without considering the need for additional dedicated resources, then these initiatives invariably fail. Delivering customer schedules on time must always remain the top priority for operational management. However, without innovative change programmes, businesses experience a steady decline in productivity resulting in an inevitable loss of business profitability.

Designing effective manufacturing systems is the only way to compete against companies that innovate through investing in manufacturing systems design, creating appropriate production facilities and training people. Why should businesses expect to beat competition by using traditional methods and outdated facilities?

Manufacturing system design process

The design of a manufacturing system comprises a number of discrete phases, starting with the business requirements and concluding with the implementation of a new supply chain designed specifically to meet the business needs. This process takes many factors into account, structuring manufacturing operations around core activities that embrace the business strategies needed to be competitive in their chosen markets, translating them into a structure capable of beating the competition.

For example, rally cars must be designed with all the technical features needed for competitiveness. However, once designed their ultimate performance depends upon attention to detail and continual development, plus tuning and an experienced team.

The same principles apply to a manufacturing process. Basic functionality must be embodied by design into the supply-chain process, developed to become world class through continual improvement and operated by a skilled team. The aim is to create an optimized total solution as opposed to a traditional approach of optimizing various subsystems.

An outline process for designing supply-chain and manufacturing systems is shown in Fig. 1.1.

Identify the need for change

- Translate the business plan requirements into operational targets that must be achieved through designing core operational processes.
- Establish a detailed market and product strategy.
- Collect information on the current supply-chain and manufacturing methods.
- Identify the critical factors influencing business performance.
- Quantify and confirm opportunities for improving operational performance.
- Verify that business plan operational performance targets are achievable.

Module/cell identification

- Identify a supply-chain policy and core product groups.

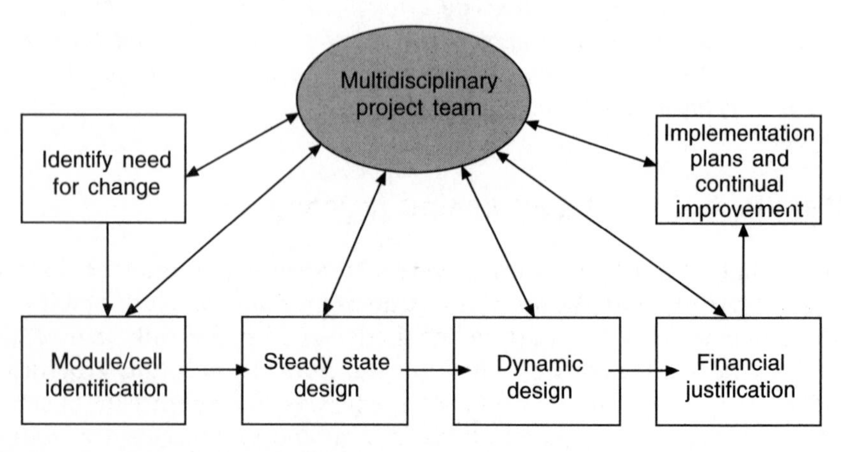

Figure 1.1 Stages in supply-chain and manufacturing systems design

- Develop a manufacturing strategy, establishing a vision and an ideal manufacturing operation.
- Confirm current performance against 'best practice' checklists for specific markets.
- Perform strategic make versus buy analysis.
- Define the manufacturing modules and cells needed in-house.
- Confirm a detailed manufacturing policy.
- Establish implementation costs and benefits, obtaining management approval and support.

Steady state design

- Determine the size of modules and cells needed to satisfy market demand.
- Design how modules/cells will operate under steady state conditions.
- Evaluate the benefits of introducing new manufacturing technology.
- Introduce Japanese manufacturing techniques to improve process effectiveness.
- Determine the job requirements for module/cell teams.
- Establish services needed to support module/cell teams.
- Define the quality system.
- Develop a personnel policy.

Dynamic design

- Modify module/cell designs to accommodate variable factors influencing operational performance.
- Identify risks and methods of limiting their potential impact upon performance.
- Select supply-chain and manufacturing control systems.
- Identify the preventive maintenance routines.
- Specify the tool management system.

Financial justification

- Develop a financial case for implementation.
- Determine the product manufacturing costs using new process-based modules/cells.
- Create a supplier integration policy.
- Identify measures of performance needed to drive continual improvement.
- Produce time-based implementation plans.
- Obtain formal investment approval and funds to implement proposals.

Implementation and continual improvement

- Install manufacturing modules/cells and commission equipment.
- Define job requirements and skills needed for the module/cell teams.
- Select module/cell teams and provide necessary training.
- Launch module/cell teams.
- Introduce continual improvement process.
- Introduce module/cell measures of performance and information systems.
- Confirm that the system delivers the business plan's financial commitments.

In the following chapters each of these items will be described in sufficient detail to provide both supply-chain management and change project teams with the tools and techniques needed to design and implement effective manufacturing systems. The size of the task must not be underestimated because design robustness is governed by analysing detailed information, forming balanced judgements and developing practical solutions, all of which take considerable time.

Project teams for manufacturing systems design

Supply-chain and manufacturing systems design is a full-time task, requiring dedicated resources from the most knowledgeable people in the business. It is imperative a business acknowledges the importance of this commitment before embarking upon a project. Under-resourced project teams fail to meet critical implementation milestones, or, worse still, implement ill-conceived solutions that cripple the business. The manufacturing systems design process described in the following chapters provides an established method of securing future prosperity but it will impact every aspect of *how things are done* within the business. Sometimes a business is already unprofitable and to redesign its fundamental manufacturing processes is a last chance of salvation. These usually prove to be very successful projects because, when faced with possible site closure, solutions tend to be radical and innovative!

Successful manufacturing systems design projects have the following features:

- Project identified in the company business plan as a 'significant change project'.
- Costs for undertaking the project agreed in the financial budget.
- Project owned by the supply-chain director or site general manager.

- Full-time project manager appointed with responsibility for delivering the programme on time and within budget.
- Core team of full-time people appointed with experience of:
  - supply-chain management;
  - product design and applications engineering;
  - manufacturing engineering;
  - manufacturing systems design;
  - current production methods;
  - manufacturing management; and
  - job design and training.
- Part-time specialist team members assigned from other business areas:
  - purchasing and supplier development;
  - customer satisfaction and quality;
  - business development;
  - aftermarket and product support;
  - finance; and
  - business systems and information technology.
- Part-time team members must commit defined periods to the project team.
- Clear terms of reference defining the scope of work and overall objectives must be understood and agreed by the local management team.
- Project plan and work packages managed using established project management techniques.
- IT-based project management and control system used for documenting plans and recording progress.
- Accessible team area created with project plans and general information on display.
- Criteria agreed for raising hazard reports.
- Senior management steering group appointed with authority to approve the team's recommendations, taking responsibility for providing the funds needed to implement proposed changes.

Project management and control

Good project management is crucial to the success of any significant manufacturing change programme. It is important a project plan with associated work packages is established at the outset. The task of accurately estimating the resources needed to complete work packages is difficult without knowing the amount of work entailed, but 'intelligent estimates' usually provide sufficient information to scope the project and determine the overall size of the team needed to complete the initial manufacturing system design. One problem facing the project manager is to determine how much effort should be directed towards particular work packages, giving essential information without generating unnecessary data. Costs at the design phase are mainly associated with funding the team responsible for establishing the new manufacturing system. This expenditure is normally relatively modest compared to implementation costs, which includes training people to acquire the broader skills needed to work in cell-based teams.

The formal project management procedures that should be adopted have been explained in detail in *Plan to Win* (Garside, 1998). For change projects these must be supported by action checklists directing work towards specific tasks fundamental to producing a robust manufacturing system. The project manager is responsible for maintaining progress towards delivering the project objectives within the budgeted expenditure limits. This is achieved by establishing a series of formal reviews, early identification of deviations to the plan and instigating corrective actions needed to recover the situation (Table 1.1).

Table 1.1 Action lists used to focus project team activities

Project team actions					
No.	*Date*	*Task*	*Person responsible*	*Date required*	*Date achieved*

Project work packages are controlled through a series of meetings, using a standard agenda and attended by designated groups:

- **Daily reviews**
  Project manager and team – held in the project office for fifteen minutes to record daily progress, identify problems and update action sheet.
- **Weekly meetings**
  Project manager, team and supply-chain managers – held in meeting room for one to two hours to identify tasks, review progress and provide guidance for resolving outstanding issues.
- **Monthly meetings**
  Site general manager, supply-chain director, project manager and site steering group – held in conference room for one to two hours with the project team, programme director and appropriate supply-chain managers to receive formal presentation and provide feedback on progress.

- **Milestone phase review**
  Senior management review – held with the business managing director, programme director and site executive team members to confirm the business objectives, support the proposed action plan and release necessary expenditure.

Each formal meeting should generate action sheets with a set of tasks; these are assigned an owner with a completion date appropriate to the agreed project plan. These provide the mechanism for keeping the project focused upon factors considered important to the business.

Manufacturing engineers

The design of supply-chain and manufacturing systems in some companies is regarded as 'a special event'. In my experience, having a core of experienced people available full time who understand how a factory works, with responsibility for continually improving the effectiveness of the supply chain, provides significant benefits for the business. They provide the foundation and expertise for the multidisciplinary project teams needed to complete supply-chain projects and deliver the financial benefits expected by the company. The manufacturing system design dictates the cost base of the supply chain and in most instances the profitability of the company, but many businesses still fail to recognize the need for these manufacturing skills within the organization.

> *It is fundamental that businesses identify who is responsible for designing how the supply chain will operate and whether they possess sufficient technical ability to create innovative solutions.*

Several leading Japanese companies report having the same number of engineers assigned to manufacturing systems design as they have to new product introduction. This is rarely the case for traditional Western businesses! The role of the present-day manufacturing engineer (similar to other jobs in manufacturing) has been expanded – in my opinion correctly – to include a range of activities previously performed by production specialists.

Areas of responsibility should include:

- Manufacturing system design, to create effective supply-chain processes.
- Machining technology needed to produce higher quality, more cost-effective components.
- Identifying core modules needed to develop and retain proprietary knowledge.

- Specifying manufacturing processes needed to ensure consistent production.
- Recommending assembly methods and levels of automation.
- Designing the logistics systems needed for transporting materials from supplier, through the factory and delivery to the customer.
- Understanding robotics, conveyor systems, handling devices and such.
- Understanding material processing and surface treatments.
- Developing alternative production techniques for producing quality parts more cost effectively.
- Implementing new production methods to achieve product cost targets.
- Working with new product introduction teams to design products for manufacture and assembly, using the installed equipment available to production.
- Ensuring product designs meet quality and cost objectives.
- Designing fixtures and tooling needed to manufacture and assemble products.
- Producing process layouts for manufacturing teams.
- Providing numerical control machine instructions for complex parts.
- Establishing tooling requirements, obtaining maximum benefit from existing ones.
- Commissioning equipment and proving processes are capable.
- Supporting continual improvement groups to implement change.
- Producing factory layouts.
- Preparing cost justification for capital investments.
- Ensuring equipment is process capable.
- Making most effective use of facilities and recommending areas for improvement.
- Ensuring all processes and equipment meet local, national and international environmental, health and safety standards.

These activities are fundamental to creating effective manufacturing systems. Manufacturing engineers with these skills are also in demand in areas such as:

- *Supply-chain and manufacturing systems design and implementation:*
  - designing the supply-chain and manufacturing system for current products to improve effectiveness;
  - designing the manufacturing system for new products;
  - commissioning manufacturing facilities; and
  - supporting the supplier base to introduce effective systems.

- *Product introduction:*
  - supporting product introduction team;
  - designing components for manufacture, assembly and cost;
  - production trials; and
  - tooling and fixture development.
- *Manufacturing technology development:*
  - developing new manufacturing techniques; and
  - evaluating alternative production technology.
- *Manufacturing operations:*
  - cost reduction on existing products;
  - 'traditional' production engineering;
  - process layouts and manufacturing documentation;
  - resolving day-to-day production problems; and
  - supporting supplier improvement programmes.

The resources needed to support normal factory operations and new product introduction should have been identified as part of the business planning process, but it is important to recognize the *additional workload* created by undertaking a significant factory design project. The precise resource requirements are relatively difficult to predict due to the complex nature of change projects, but success depends upon collecting and analysing information on all aspects of the manufacturing process. This takes considerable time and effort if it is to provide meaningful results. In my opinion, those external consultants promoting an impression this work can be completed very rapidly have greatly underestimated the amount of work involved in designing factories. They generally suggest only partial solutions, often failing to gain support of the people who ultimately must implement the proposals. The team owning the project and taking responsibility for effecting the change *must* be members of the local management team and engineers from within the business. If external consultants are *then* used to support these people, a valuable dimension is added to the project team, usually leading to further innovative ideas.

Effective manufacturing systems

Effective manufacturing systems are the foundation of Japanese modular, lean production facilities. The achievement of world class operations capable of competing in international markets is a combination of good manufacturing systems design, supported by continual improvement programmes that consistently refine the processes, eliminating all types of waste. The Japanese have taught Western companies many new techniques for improving manufacturing operations; these have been widely adopted by the motor industry and major component suppliers. The starting point for making the transformation is to understand the difference between the *efficient* production systems established by traditional Western companies and *effective* systems that were introduced, for example, by the Japanese car manufacturers (Table 1.2)

Table 1.2 Efficient versus effective systems

Efficient	*Effective*
Conventional plant Functional layout Slow changeovers	Flexible partial automation Cells layout for materials to flow Rapid changeovers
Large batches, long runs	Cost-effective small batches
Seeks economies of scale Maximizes machine utilization	Overproduction avoided Manufacture quantities required
Planned inventory buffers	Minimized levels of inventory
Long lead times	Collapsed lead times
Make to forecast	Make to order
Complex production control	Simple production control
Difficult operation to manage	Easier operation Manageable production
High overhead support	Low overheads
High production costs Customer insensitive Poor ownership of quality Poor schedule adherence	Low production costs Responsive product availability Good ownership of quality Good schedule adherence

The skills developed in Japan to achieve competitive manufacturing strategies were:

- Designing economic, high product variety manufacturing systems.
- Implementing process capable, low cost, high quality manufacturing facilities.
- Perfecting flexible manufacturing systems, integrating manufacturing methods and production technology to produce cost-effective products within short delivery lead times.

This required the design and implementation of simple, focused manufacturing modules with short changeover times, capable processes and visible control systems. These features would be specifically designed to meet specific customer requirements (Toyota, Honda, Denso, Sumitomo).

Reviewing benchmark companies who have introduced new manufacturing systems based upon simple structures designed to create an even flow of work through manufacturing modules shows a consistently achieved superior performance on all the high level performance indicators. A general consensus would indicate that a 30 per cent performance gap still exists between many Western and Japanese companies, although by implementing similar manufacturing philosophies the West now has some excellent examples of lean factories that can compete directly with the best in the world. (Emerson Electric, Milican Inc., Lucas Aerospace – Cargo Systems, Johnson Controls).

In order to develop modular, lean manufacturing facilities it must be understood that:

- Lean manufacturing is a systems approach that designs tailored manufacturing modules to meet the needs of the customers, making a range of products in the sequence required to meet customer promise dates.
- The business is organized around a number of fundamental business processes with a structure that reduces large specialist departments and replaces them with integrated multidisciplinary modules and cells that take responsibility for a complete product, component or process.
- The manufacturing design process must consider the following critical factors:

  - range and volume of products to be manufactured;
  - requirements and demands on the complete supply chain;
  - process flow, routings and customer lead times;
  - capability of the processes and measuring systems;
  - changeover times and flexibility needed to manage variety;
  - bottleneck processes and limitations on capacity;
  - methods of monitoring and controlling factory processes;
  - quality procedures and product conformance;
  - elimination of waste through continual improvement;
  - synchronization of material flow and the use of materials handling; and
  - job structures and the training of people.

- Supply-chain and manufacturing systems must be designed using a multidisciplinary project team with the skills and experience to restructure the overall supply-chain process.

Company business plan

Many businesses will have developed a business plan, providing information on the future direction, objectives and financial targets for the company. I believe a robust business planning process is a crucial element in creating a *winning* business, and is the most important document prepared by the management team. This plan must be consulted throughout the supply-chain system design process and all actions must be consistent with meeting the objectives identified in this important and strategic document.

An example of a typical business plan and the type of information required include:

The introduction

Summary of the business and its position within the industry:

- Present market and expected trends.
- Outline of the business position in the market.
- Major issues that have been identified in the business plan.
- Summary of actions needed to achieve the financial commitments.

The key issues facing the business and significant features related to new and existing products are shown in Tables 1.3 and 1.4.

Table 1.3 Example of market characteristics

Market characteristics	
Strong growth in US/European markets	Dominated by US industry
Recovery in aftermarket demand	Captive aftermarket
Controlled by five key suppliers	Technical barriers to entry
Market has overcapacity	Weaker companies will not survive

Table 1.4 Business strengths and weaknesses

Strengths	*Weaknesses*
Good market position	High manufacturing cost structure
Strong technology	Engineering not customer-focused
International presence	Functional organization
Broad product base	
Good next generation technology	

Summary of business objectives (example)

- Achieve world class cost and asset performance through implementing effective manufacturing process.
- Consolidate manufacturing operations into core technologies and increase overall asset utilization.
- Delight customers through exceeding their expectations.
- Achieve product technology leadership.
- Organize the business around key business processes.
- Strengthen presence in overseas regions.

Market overview and sales plan

Market overview provides a global projection of the major segments important to the business, the range of products within the business portfolio and level of *sales, gross margin and profit* expected by major product lines over the plan period.

The business plan should provide detailed information concerning:

- Major market segments.
- Prime factors that drive the market.
- Movement in specific market segments.
- Specific events that will impact the market.
- New legislation and social trends.
- Impact of new technology.
- Influence of competitors on the value chain.
- Movement in exchange rates and monetary policy.

Information available may also include:

- Movement in the global economy and major markets.
- Predicted growth in overall product demand by territory.
- Related economic factors that influence the potential market (for example, new building starts, increased wage rates, increased car sales, expected aircraft orders).
- New technology that will impact the market.
- Mergers and industrial consolidation.

Market segmentation and projected trends
Definition of the key market segments, together with information illustrating the relative size and growth of each element within the segment, including the impact these factors may have on business performance.

Current and projected market share
Comparison of the business performance against competitors, position

of existing products and those with growth potential. Also, which products generate the majority of sales and profit, within the market segments they serve.

Customer characteristics and profiles

Analysis of the customer base original equipment manufacturers, aftermarket customers, governments, or other manufacturers in the value chain (Table 1.5). Different strategies for generating sales in each of these markets, understanding the primary and secondary qualifying requirements for obtaining sales with particular customer types. Reasons for obtaining sales in particular market segments:

- Price.
- Total cost of ownership.
- On time delivery.
- Legal requirements.
- Place of manufacture.

Table 1.5 Customer profiles

Prime OEM	*Position*	*Sales*	*Importance to company*

Government departments	*Position*	*Sales*	*Importance to company*

End-users	*Position*	*Sales*	*Importance to company*

Competitor profiles

Assessment of major competitors, comparing their position in the market and financial strength (Table 1.6). Summary of recent actions illustrating strategic direction or increase in market share through the introduction of new technology or business acquisition.

Table 1.6 Competitor profiles

Competitor	*Position (market/financial)*	*Actions (recent/anticipated)*	*Company response*

Product introduction

Analysis of current product base, the life cycle of major product lines and their position in that life cycle. The core technologies embodied within the product base and the potential for further commercial exploitation in new products. The business plan should also state the *core technologies* to be developed and retained, if the company is to secure a dominant business presence in its chosen markets.

Product introduction plan
This structures business opportunities and identifies:

- Actual customers and subsequent customer chains.
- New projects, major customers and potential sales value.
- Products/components/subsystems to be supplied.
- Partners and collaborators in projects.
- Dates of key project milestones for new programmes.
- Potential value of sales to the company.

Project summaries
Description of significant product introduction projects and the impact they will have upon supply-chain processes.

Current project status
Relative position of development projects, in the product introduction process confirming where they are in the development cycle (see Fig. 1.2).

Supply-chain management

Management plans for improving supply-chain operational performance are based upon current manufacturing process knowledge and future business strategy. These factors should have been addressed:

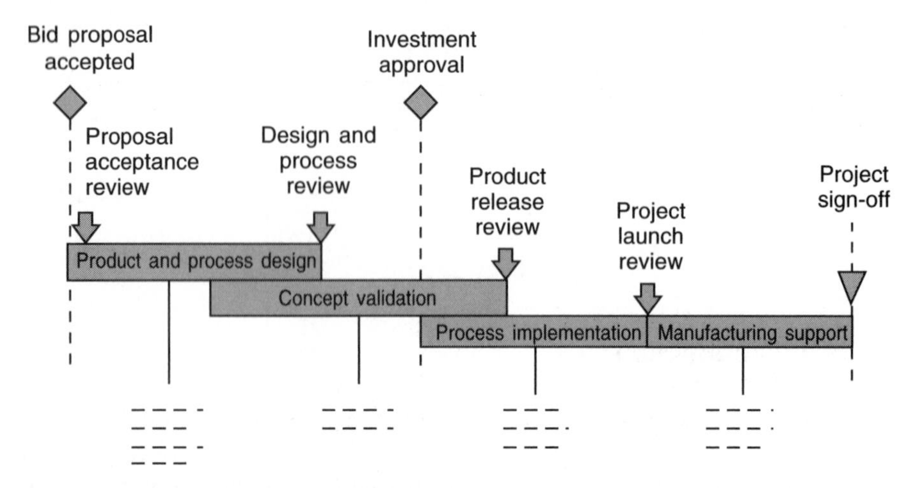

Figure 1.2 Status of projects in the development cycle

- Market territories served.
- Geographical location of facilities.
- State of the manufacturing facilities and capacity available.
- Level of investment available for capital equipment.
- Robustness of the product base.
- Skill of the workforce.
- Availability of training in core skills.
- Strength of the supply base.

Present situation and current status of supply-chain projects

These may include items such as:

- Strategic make versus buy, to focus upon core components.
- Consolidation of manufacturing facilities.
- Restructuring operations, focusing upon assembly and test.
- Transfer of core machining to more appropriate locations.
- Opening new manufacturing facilities to secure expanding market opportunities.
- Restructuring the supply chain, with investment in new manufacturing processes to protect core technologies.

Operational improvement

The strategic ambition and short-term supply-chain operational improvements fundamental to future profitability, ensuring a cost structure that allows the business to survive:

- Location strategy.
- Key elements of the supply-chain process that must be redesigned to enhance operational performance.
- Internal supply-chain policies, focusing upon manufacturing core components, assembly, test and aftermarket support.

Different manufacturing systems are required to produce various component types. The business plan should provide guidelines on the appropriate manufacturing modules and associated cells needed to create a flow of work in a rhythm that paces the factory. It should also define the key elements of the *supply-chain process* to be considered when developing supply-chain and manufacturing systems. These include:

- *Make versus buy analysis,* establishing those items to be bought out and those manufactured in-house. This should be a strategic decision and not a short-term expedient to reduce product costs.
- *Supply-chain structures,* identifying runner, repeater and stranger products, with separate manufacturing systems for servicing different market/volume requirements.
- *Modular organization structures,* identifying natural working groups who take full responsibility for all activities needed for completing a component or process.
- *Master production scheduling procedures,* implemented at three levels to establish the demand across the factory and with suppliers.
- *Demand driven material flow systems,* pulling components through into assembly, shortening lead times, reducing stock and work in progress, making problems visible and removing tolerance of failure.
- *Changeover times reduction,* introducing specialist equipment and procedures to support the changeover process.
- *Continuous improvement* through a relentless drive to eliminate all forms of waste.
- *Process capability of machine tools,* measuring equipment, materials processing, and assembly and test systems, through the application of statistical methods and experimental techniques to resolve problems.
- *Documentation for all processes and procedures,* making them user-friendly, supported by graphical/pictorial representation to aid understanding.
- *Materials handling,* introducing higher cleanliness levels and using dedicated containers to protect components as they move through the supply chain.
- *Modern IT systems* for manufacturing planning and control, giving real-time information on workflow and manufacturing priorities.

Selecting the areas that provide the greatest overall business benefit is crucial for supply-chain and manufacturing systems design programmes. Deficiencies in these areas can be identified using checklists to highlight gaps against known best practice. However, ranking them and assessing relative returns requires considerable management judgement.

Purchasing

Many businesses spend over 60 per cent of revenues on purchased items. The product introduction process requires the greatest purchasing expertise, establishing a competitive supplier base for future products, but the continual drive for cost reduction will only be achieved through consistently seeking ways of reducing the overall costs of raw materials, consumable items and services. The plan should identify categories of components with the greatest spend profiles, or most significant supplier problems, providing supplier listings for analysis (Table 1.7).

Table 1.7 Supplier information

Commodity	*Spend £ m*	*Number of suppliers*	*Number of suppliers – 80% of spend*	*Target cost savings %*	*Performance rating – cost quality delivery*

Distribution and aftermarket

Successful methods of product distribution lead to significant competitor advantage. New ways of reaching the end customer are being exploited with changing customer-purchasing habits. So, the route products take to market is a critical aspect of the overall supply-chain process. The logistics systems required for new product and aftermarket parts might be different, and it is important to understand the dynamics and customer requirements of both market segments.

The relative importance of the original equipment (OE) and aftermarket business sectors is dependent upon the type of products, their position in the value chain and the proportion of sales and profit generated directly through the aftermarket business. However, it is important to understand the profitability of service parts, because while resale prices may be considerably higher, reported manufacturing and distribution costs may not reflect the invisible cost of disruption within the supply-chain process. In some industries, distribution and aftermarket support is organized as a separate business. Consequently, the team must review:

- Role of distribution and aftermarket in the overall business.
- Market changes influencing the different market segments.
- Specific actions being taken by customers within the value chain.
- Differing requirements of original equipment and service part customers.
- Position of the company in the industry, relative market shares for original equipment, aftermarket, spares and repairs.
- Relationship with major customers, stocking policies and other key factors that influence sales.

The crucial factor in the distribution and aftermarket process is to consistently achieve industry standard customer service levels. Delivery promise dates must always be met to secure ongoing business; the company will not ultimately survive unless it meets expected customer performance levels.

Financial results

The finance section of the plan shows past performance, actual and budget for the current year, and projections over the plan period with benchmark targets:

Financial summary

- Sales.
- Trading profit.
- Restructuring and major change project costs.
- Redundancy and closure costs.

  ⇒ *Profit before interest and tax*
  ⇒ *Operating cash flow*

- Opening cash balance.
- Closing cash balance.
- Employees – *Year end*
  Average
- Payroll costs.

Key ratios

- Return on sales.
- Capital turnover ratio.
- Return on capital employed.
- Sales/employee.
- Added value per £ payroll cost.
- Payroll costs as percentage of sales.

Engineering costs

- Engineering funding.

  ⇒ *Net engineering costs*

Capital expenditure

For new products, manufacturing capacity, quality, safety/environmental, productivity, cost reduction, replacement equipment, plus other items including tooling, information technology, vehicles, land and buildings, etc.

- Capital expenditure.
- Depreciation.

Stock analysis and stock to sales ratios

- Value of stock: analysis by raw materials, work in progress, maintenance, consumable materials, finished goods and replacement parts.

Financial data presented in the plan may vary, but it should provide details on current performance and future financial commitments. It also forms a basis for comparing the impact various options have on financial performance.

Business strategies and actions

Business strategies and policies required for securing future business viability.

Strategy development

- *The position of the business in the industry*, comparing the size of the business with other companies, its ability to compete effectively and relationship with competitors.

- *Activities across the industry* that are impacting available capacity, reasons for expansion or consolidation, the cost pressures within the industry, potential acquisitions, divestments or alliances, and actions being taken by competitors to secure commercial advantage.
- *Strengths, weaknesses, opportunities and threats* analysis for the business, demonstrating the potential for growth and actions that must be taken to retain stakeholder value.
- *Commentary by major or potential market territory* on the investments made, and action the company will take in relation to political and economic factors in certain areas of the world. Consideration should also be given to the structure of industry, location of manufacturing sites, status of local manufacturers and the reputation of the business.
- *List of potential acquisitions divestments and alliances*, giving the name of parent company, company name, sale revenues and making the business case.
- *Vision of the business* and its position within the industry, describing the critical actions necessary to accurately position it.
- *Role* of the business within the company portfolio.
- *Strategy and goals* that have been established for the business.
- *Policies* that must be implemented to sustain a viable business and satisfy its role.
- *Critical success factors* that must be achieved to meet the business challenges.
- *Business performance objectives* that must be delivered, including financial commitments and significant projects.
- *Tactical actions* necessary to be a successful management team.

Detailed information on how to compile a business plan is available in *Plan to Win* (Garside, 1998), which identifies the key elements of a business planning process.

Setting business targets

The supply-chain and manufacturing systems design process can be divided into a number of phases. The first task is to confirm *business targets* and objectives that encapsulate management direction and meet business aspirations. The formal business plan prepared by the management team is the most reliable information available and must be used as the foundation for designing the supply-chain and manufacturing systems. If the information identified in the previous section has not been collated into a formal document, then the project team's first task is to collect the company information needed to give the supply-chain redesign project strategic direction.

Once the business targets have been established they should be compared with benchmark performance figures for particular market sectors, providing competitive goals and indication of any performance gaps that may exist against major competitors. High-level performance targets must be determined for the business. These are used to validate alternative solutions by assessing their likely impact on selected key performance indicators (Table 1.8).

Table 1.8 Key performance indicators

Financial	*Current*	*Benchmark*	*Target*
Return on sales			
Return on capital employed			
Capital turnover ratio			
Sales per employee			
Added value per unit of pay			
Stock to sales ratio			
Customer Satisfaction	*Current*	*Benchmark*	*Target*
Customer satisfaction index			
Cost associated with quality			
Number of defects returned from service			
Adherence to customer delivery schedules			
Lead times needed to support deliveries			
Manufacturing effectiveness	*Current*	*Benchmark*	*Target*
Manufacturing costs			
Inventory levels			
Cost index for bought-out items			
Manufacturing and supplier lead times			
Overhead savings			

The process for setting business targets is shown in Chart 1.1.

These business plan objectives and performance targets are then used as the driver for designing effective supply-chain and manufacturing systems. An outline of the overall process is show in Fig. 1.3, illustrating how the different factors must be considered when designing an integrated system.

The more effort management spends developing a robust business plan, the easier it is to define manufacturing requirements. The business plan as described provides the overall direction, but will not provide the

Chart 1.1 Defining business targets

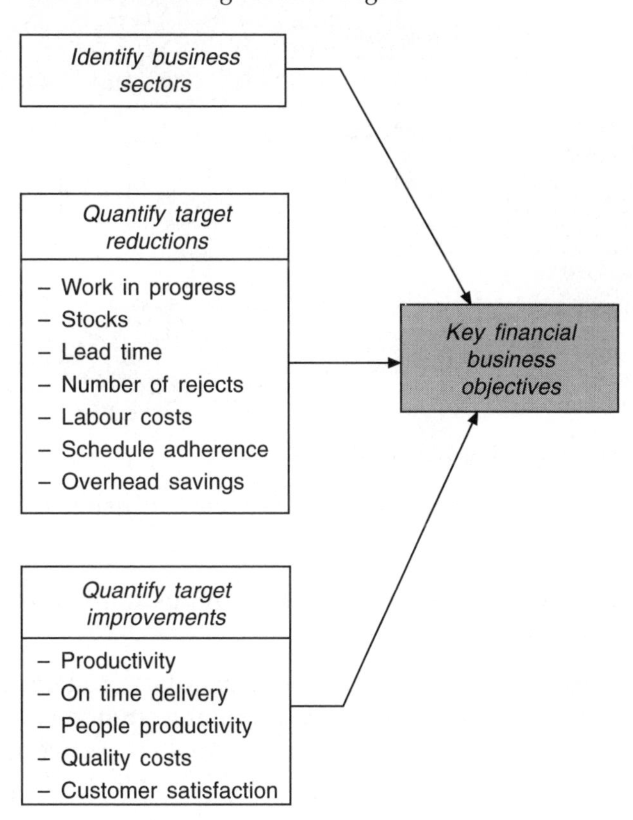

detailed information needed for developing an effective manufacturing system.

Detailed market and product strategies

The next phase of the task is to structure more detailed marketing and product strategies with the specific aim of determining the current and future supply-chain requirements. This analysis must take full account of the operational and financial business objectives identified in Chart 1.2. It must also consider the key market segments and associated product groups. If necessary, an analysis must be undertaken on significant product groupings, without allowing the task to become unwieldy and unproductive.

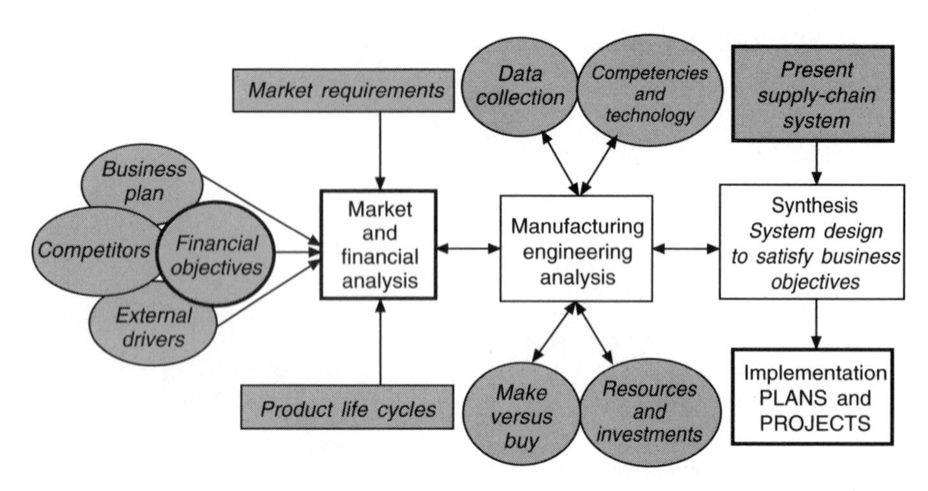

Figure 1.3 Manufacturing systems design process

This information requires detailed evaluation using the following techniques:

Supply-chain – strengths, weaknesses, opportunities and threats analysis (SWOT)

The SWOT provides the supply-chain and manufacturing understanding needed to start the design process, by examining strengths, weaknesses, opportunities and threats from both an *internal perspective* and *external appraisal*. It is usually undertaken by a multidisciplinary team who:

- Brainstorm various aspects of the business.
- Analyse and clarify critical elements.
- Determine factors underlying strategic issues.
- Rank factors in order of importance to the business and its customers.

Tables 1.9 and 1.10 show examples of factors that might be considered.

Market analysis

The two most important aspects from the market analysis are determining the *range of products* to be manufactured and assessing *future volume* requirements. This information is critical for establishing production capacities and the level of investment that can be justified installing alternative manufacturing methods. These requirements should have formed a key element of the business plan, but they now need further detailed evaluation to quantify:

Product life cycles
All products pass through a number of stages:

Chart 1.2 Market and financial analysis

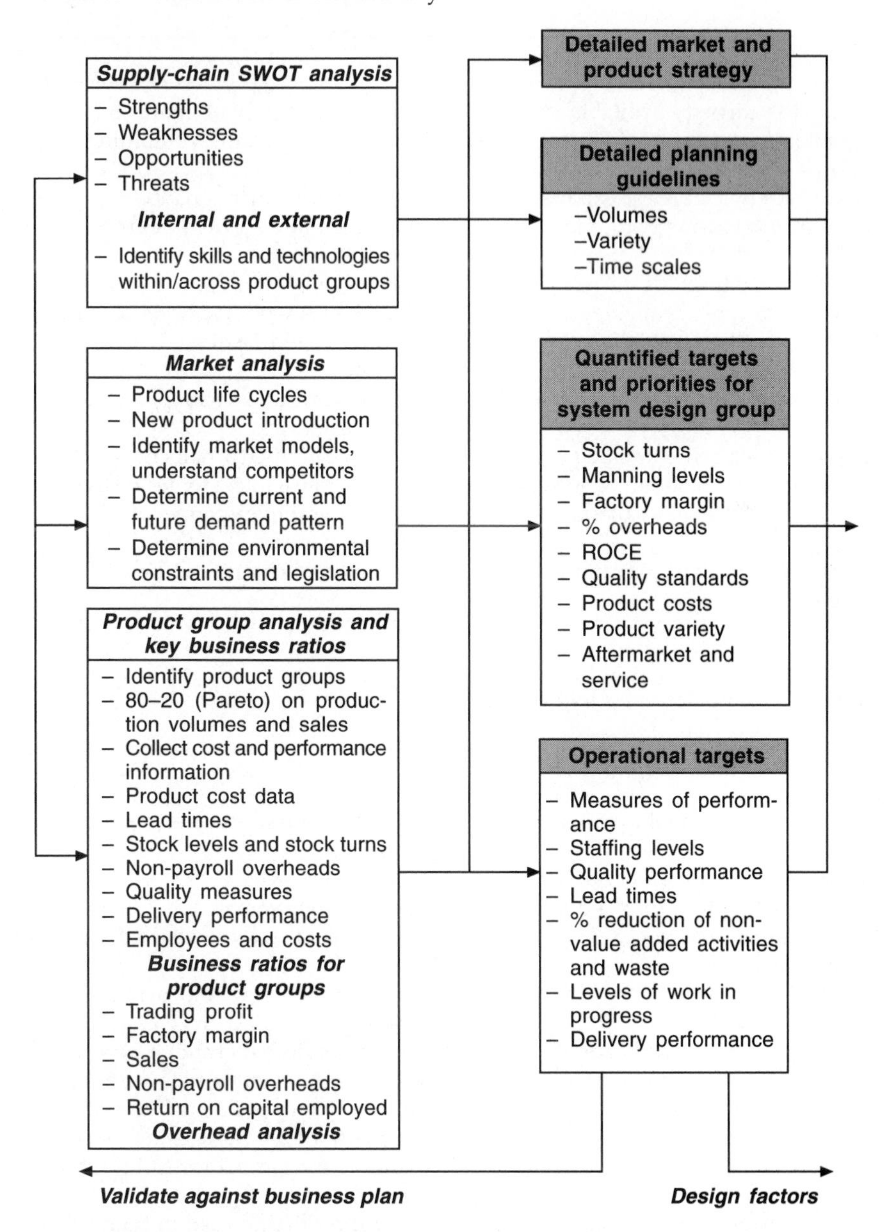

- *New products entering the market* – have a low market share with potential for growth. The growth rate and capacity have to be determined.
- *Current products with further potential for growth* – the challenge is to identify market opportunities and manufacturing capacity needed to exploit their full potential.

Table 1.9 Supply-chain strengths and weaknesses

Strengths	*Weaknesses*
Production capacity available	Plant old and in poor condition
Capable of producing wide product range	Processes not capable of holding tolerances
Floor space available for expansion	No preventive maintenance
Materials processing equipment on site	Planning and control systems inadequate
Tool room facilities	Restrictive working practices
Technical support available	Several sites with duplicate facilities
Process development team established	Traditional, old fashioned factory layout
Material laboratory and expertise	Large organization with excessive overheads
Assembly area self-contained	Poor supplier performance
Test facilities automated – spare capacity	Slow response to customer requests
Good range of plant available	Lack of internal measures of performance
Capable of supplying finished components	Poor internal communication

Table 1.10 Opportunities and threats

Opportunities	*Threats*
Utilize or close available capacity	Competitors introduce lower cost products
Invest in new plant and equipment	Manufacturing moves off-shore
Change tooling to improve quality	Level of investment cannot be justified
Introduce team-based cell manufacturing	Management lack the vision for change
Rationalize facilities	Competitors merge to provide cost savings
Remove excessive overheads	Customers find alternative suppliers
Introduce process capability studies	Major volume with single customer
Simplify material flow	Change in management direction
Reduce changeover times	New technology changes the business
Train people to become multi-skilled	Acquired by competitor or investment house
Introduce improvement programmes	Run for profit and cash to fund other ventures
Redesign the supply-chain systems	People will not accept altered responsibilities
Establish good scheduling systems	Recurring quality problems
Improve the supplier base	Loss of customer confidence

- *Established products* – generate sales and profits through regular demand from identified markets. The task is to accurately predict when the sales volume will fall and the manufacturing process becomes uneconomic using existing manufacturing routes.
- *Geriatric products leaving the market* – required in lower production volumes and service parts for the aftermarket.

The task is to analyse the full spectrum of products over the planned time horizons and to predict annual volumes for each significant product group. The task is complicated by having to take account of *outside factors* which may influence sales, such as:

- New technology products entering the market.
- Changes in the products' designs to meet performance criteria.
- Strategic sourcing and pricing.
- Loss of sales to competitor products.
- Substitution sales to other products in the product portfolio.
- Changes in legislation and customer expectation.
- Impact of exchange rates and off-shore manufacture.
- Trends and changes in fashion.
- The need for local manufacturing capability.

The less predictable the market requirements, the more flexible the manufacturing system needs to be in order to accommodate inevitable changes in demand.

Techniques are available to improve forecasting accuracy, through analysing situations, making informed assessments and taking management action to sustain sales volumes. These include:

- Ranking customers, rating the chance of obtaining sales.
- Reviewing the relationship between sales volumes and past events.
- Examining seasonal demand variations and changes in product mix.
- Using price as a mechanism to secure projected volumes or level demand.
- Determining high and low projections, linked to probability of events.
- Examining relationships between market share and product variety.
- Evaluating price implications resulting from increased product variety.
- Reviewing past events to predict future demand.

These techniques should be applied whenever possible to refine and confirm the production requirements because the result of these projections will have a direct influence on the final supply-chain and manufacturing systems design and related product cost structures.

Forecasting is an imprecise process, but no matter how inaccurate, it is far preferable to have agreed sales projections than merely attempting to react to unplanned events.

Questions to be addressed because of their direct implications upon supply-chain processes:

- How reliable and accurate is the market forecast?
- Should investments be customer or product line specific?
- Can the system be made more flexible, aiming at more general markets?
- What are the implications for delivering an improved manufacturing performance? Will it:

  - enhance product quality?
  - reduce product costs leading to increased demand?
  - open new markets for the product?
  - allow further investment for developing the product range?
  - improve profitability?
  - provide greater security for the workforce?
  - remove competition from the marketplace?
  - allow expansion into new market territories?

- How effective will the manufacturing unit be at producing the projected volumes and mix of products?
- Can the facilities be used for other product lines if the forecast demand is not achieved?
- Will a change in product base result in expensive retooling?
- Will the equipment be suitable to manufacture new and existing products?
- Is the fabric of the building and facilities conducive to further investment?
- Does the company have similar facilities that could be consolidated to remove overheads and fixed costs?
- Does the workforce have the necessary skills and will it support the proposed changes in working practices?
- Can the production facilities accommodate runner, repeater and stranger products?

Product group analysis and key business ratios

Financial information and production data for each of the product groups included in the market analysis should be compiled and entered in standard PC spreadsheets, which can be used to manipulate data into meaningful information. The type of analysis required to structure the data and make informed business decisions includes the product groups categorized as:

- *Runners* – products in regular production with all manufacturing planning information, NC programmes, tooling, fixtures, heat treatment, assembly instructions and test data including quality procedures fully documented and available to production.
- *Repeaters* – products manufactured at regular intervals, but not fully tooled for volume production. The documentation requires skilled operators to take responsibility for ensuring consistent product quality.
- *Strangers* – items required for product and process development or low volume aftermarket components needed to repair geriatric products.

A distinction is made between these categories of product because they require different manufacturing systems suited to particular production volumes and range of product variety. This differentiation also allows associated manufacturing and overhead costs to be more accurately attributed to particular product groups, reflecting actual production costs.

Information required for each product group includes:

- Total value of sales.
- Number of customers.
- List of the 20 per cent of customers that generate 80 per cent of the sales (Pareto) – ranked in order, giving details of:
  - overall account value;
  - variety of products in the group;
  - volume of each product type;
  - sales value attributed to each type;
  - category of product – runner, repeater or stranger;
  - level of aftermarket sales; and
  - position on the product life cycle curve.
- Manufacturing costs attributable to particular product groups:
  - raw material costs;
  - value of bought-out components;
  - direct labour costs;
  - factory overheads;
  - production costs of core components;
  - heat treatment and processing;
  - assembly;
  - test and qualification;

  - specialist facilities;
  - maintenance and repair; and
  - consumable items (including tooling).

- Manufacturing costs showing actual against standard costs.
- Parameters that differentiate products in the range.
- Lead times for product groups:
  - customers placing an order to actual delivery;
  - lead times from release into the factory to customer delivery;
  - average lead time quoted to the customer;
  - time spent adding value in production;
  - lead time on the computer system for manufacturing products;
  - lead time on the computer system for the supply of key components;
  - lead time for suppliers delivering critical components; and
  - accepted market lead time.
- Stock to sales ratio.
- Stock value and percentage of stock in:
  - raw materials;
  - bought-out components;
  - work in progress;
  - finished goods; and
  - service parts.
- Non-payroll overheads attributable to the product range:
  - depreciation;
  - operating leases for plant and equipment;
  - rent, rates and local taxes;
  - factory facilities and maintenance;
  - gas, electricity, fuel oil and chemical treatments;
  - waste and effluent disposal;
  - heat treatment and surface modification;
  - transport and distribution;
  - selling and advertising expenses;
  - consumable materials;
  - security;
  - insurance premiums;
  - computing costs and data management charges;
  - packaging and storage;
  - on-site customer support;
  - cleaning; and
  - protective clothing.

- Delivery performance:
  - percentage achievement of the master production schedule;
  - percentage achievement of original equipment deliveries to customer requirement dates;
  - percentage achievement of spares to customer request date; and
  - percentage purchased items delivered on time to the master production schedule.
- Number of people employed on:
  - direct work;
  - supporting operations; and
  - supervision, management and staff.
- Payroll costs for:
  - direct workers;
  - support people;
  - supervision; managers and staff; and
  - overtime payments.

In most businesses detailed information such as this will *not* be readily available. A balance is required between obtaining sufficient detail enabling meaningful decisions to be taken and burdening people with an onerous data collection task. It must be remembered that these figures are not for the audited accounts and 'best estimates' are generally adequate.

Even informed guesses are better than having no information on which to base decisions.

The next stage is to determine key *business factors and ratios* for each product group:

- Product group original equipment sales to total business sales.
- Product line factory margins for original equipment.
- Factory margin as percentage of product group sales.
- Product group trading profit.
- Product group profit as percentage of total sales.
- Service parts sales value.
- Profit on service parts as percentage of product group sales.
- Non-payroll overheads attributable to the product group.
- Value of stocks.
- Stock to sales ratio.
- Capital employed.
- Return on capital employed.

A manufacturing functional analysis, selecting those parameters that should provide an insight into the main cost drivers for product groups include:

- percentage cost of materials and purchased components to sales.
- percentage cost of direct labour to sales.
- percentage total employee costs to sales.
- percentage non-payroll overhead costs to sales.
- percentage cost of quality to profit.
- Added value per unit of pay.
- Sales per employee.
- Added value per employee.
- Energy costs as percentage of sales.

Manufacturing data

The next stage is to review product manufacturing, supply-chain, quality and production information that must be compiled to support the manufacturing systems design process. This information also needs to be collated into a spreadsheet allowing similar analysis and consolidation.

Manufacturing data collection

Chart 1.3 shows the data to be collected for each product group.

Product manufacturing data

- Expanded bill of materials for components in each particular product group, showing parts at various stages of manufacture and at different levels on the bills of material. These are used to:
  - confirm critical activities at each level in the process;
  - identify opportunities for combining processes and removing levels;
  - establish common parts used across different product ranges;
  - identify processes to be performed at each workstation or cell;
  - define manufacturing characteristics needed to meet the product performance specifications; and
  - identify core processes critical to maintaining product technology/ manufacturing competitive advantage.
- Product routings – plotting the path critical components take through the factory. These usually show that components travel excessive

Chart 1.3 Current supply-chain information

Internal manufacturing data for each product group

Product – manufacturing data

- Bill of Materials at different manufacturing levels
- Analysis of common parts across product families
- Production characteristics of the key components

Process routes

- Flowchart of component movement around the factory
- Common routes for families of components
- Non-value added activities in the production process
- Work content by work centre or process cell

Machines and process capacity

- Capacity of facilities and bottleneck processes
- Availability of plant and equipment
- Time lost through incapable processes, unplanned events
- Changeover times on bottleneck processes
- Shift patterns, hours worked and available capacity

Quality assurance

- Customer satisfaction as perceived by the customer
- Cost of quality for both internal and external failure
- Technical compliance – in production and with customers
- Commercial compliance
- Process capability of key machines and measuring systems

Supplier performance

- Number of suppliers by commodity group
- Annual spend by commodity group
- Performance examining cost, quality and delivery

Functions and facilities

→ Information for taking strategic make versus buy decision

distances, losing ownership on the way and increasing the likelihood of unacceptable component damage through 'nicks and dings'. These routings are analysed for:

- showing common routings that exist for a range of products;
- demonstrating the long distances parts travel round the factory;
- determining time-wasted moving parts;
- evaluating non-productive time that components spend waiting for machines to become available.

- Non-value added activities undertaken as part of the current process are analysed to determine how they might be eliminated in the new manufacturing system. These include:
  - transport and movement of materials throughout the supply chain;
  - stock and work in progress at different stages of manufacture;
  - overproduction of products and components ahead of requirements;
  - waiting for materials, tooling, work instructions and equipment to become available;
  - inefficient production operations due to poor procedures and tooling;
  - operators having to walk long distances between workstations;
  - defective parts and inspection routings not performed as part of the process; and
  - equipment failures and unplanned stoppages.
- Work content for each product based upon time estimates for:
  - adding value to parts;
  - performing work requiring additional manual input;
  - machining components;
  - working bottleneck or critical machines;
  - setting up machines and process;
  - performing heat treatment and processing;
  - assembling components;
  - verifying component quality and final test; and
  - packing and shipping products.
- Production capacity (based on existing facilities, using current manning levels and shift patterns) available to:
  - manufacture products within the product group;
  - perform bottleneck processes at current level of utilization;
  - operate the key machines;
- Maximum production capacity:
  - operating bottleneck process 24 hours a day, 7 days a week;
  - running key machines 24 hours a day, 7 days a week.
- Lost capacity from:
  - production time needed for preventive maintenance on key processes;
  - poor reliability of bottleneck processes and key machines;
  - unplanned events and equipment breakdown;

  - an inability to manufacture consistently due to the inadequate process capability of machines and measuring systems; and
  - machine changeovers and resetting critical parameters on the key processes.
- Overall estimate of the maximum machine capacity and production volumes available from production based upon current shift patterns and 24 hours a day, 7 days a week working.

Quality assurance and the cost of quality
The existing quality system should be reviewed and the fundamental principles for achieving good quality products critically assessed. Information should be collected on the following:

- Customer satisfaction:
  - measures related to current customer reports on quality, price and delivery performance;
  - customer perceptions based upon assessing:
    - ◊ communications with the customer;
    - ◊ commercial competitiveness of the products;
    - ◊ equipment performance in service;
    - ◊ technical status of products; and
    - ◊ support from the business in responding to customer problems.
- Cost of quality:
  - failure costs due to not performing the work 'right first time';
  - internal costs associated with scrap and rectification including both labour and materials;
  - external warranty claims and costs of rectification including product substitution programmes;
  - quality assurance costs for operating the system, including calibration, process capability, preparation of documentation and training; and
  - impact on profit due to its overall cost.
- Technical compliance:
  - number of units rejected by the customer over number of units dispatched;
  - product conformance determining the number of concessions raised compared to the number of units shipped;
  - product data integrity monitoring the outstanding manuals and

instruction documents needing modification compared to the number in circulation; and
  - product design changes introduced to assist production.

- Commercial compliance:
  - delivery of items on time to meet agreed customer schedules;
  - delivery lead times expressing the time between receiving an order and the time taken to deliver the product to the customer.

- Process capability:
  - machine capability giving the percentage of machines known to have a process capability index of 1.33 or greater compared to the general machine population;
  - measurement system capability giving the percentage of systems known to be accurate to less than 10 per cent of the allowable product key characteristic variation.

Supplier performance

The total expenditure with outside component and commodity suppliers should be segregated into commodity groups, which identify categories of components showing the greatest spend profile. These should be reviewed using the following parameters:

- Commodity type.
- Number of suppliers in the commodity group.
- Annual spend by commodity group and key suppliers.
- Number of suppliers taking 80 per cent of the annual spend.
- Performance rating for the group examining cost, quality and delivery.
- Significant expenditure profiles with suppliers of oil, electricity, gas, services and utilities.

Factory operations

Internal supply-chain operations should be documented and reviewed based upon current product groups to provide a 'footprint' of the factory. The information collated should include:

- Details on the ownership of the facility, annual rent and terms of lease.
- Area of the site, covered area, opportunity for expansion, other facilities and any restrictions on use.
- Area of offices, production facilities, workshops and warehouses.
- Space in current use.
- Age of buildings and condition of facilities.

- Main activities conducted on site.
- Core processes key to the long-term profitability of the business.
- Major customers and locations.
- Customer's critical purchasing factors.
- Justification for maintaining the site.
- Significant investments, plant and equipment, facilities and such.
- Cost index of the site.
- Links to other businesses in the group.
- Critical items of capital equipment.
- Age of key plant and equipment.
- Numbers of people employed on site in customer development, product introduction, supply chain, distribution, finance, customer satisfaction, quality and programmes.
- Skill level of the workforce.
- Average length of service and age profile.
- Specialist skills that are crucial to the operation.
- Current shift patterns and levels of overtime.
- Unions represented on site and strength of membership.
- Supplier networks and established partnerships.
- Availability of skilled labour and resources.
- Training available and support for introducing new skills from existing workforce, local colleges and universities.
- Status of the local area for obtaining investment grants and government support.
- Senior management attitude towards the location.

Controlling the scope of data collection

Collecting this information can develop into an unwieldy task and *unless carefully controlled, grows into a significant non-value added activity!* The objective of data collection is to identify those factors having the greatest impact upon business performance and provide a quantified assessment of their importance. The relative order of magnitude for the different factors is more critical than precise detailed figures. Experience has shown that businesses possessing the majority of this information and having it readily available tend to be better managed, because in practice this information is fundamental to their successful operation. One method of keeping the task in perspective is to set a realistic time scale for collecting base information, with agreement to seek additional data if the need arises. However, effort must be directed towards evaluating a spectrum of supply-chain activities to assess their significance, not simply focusing upon those traditionally considered important.

Project assessment and review

The data and information collected usually provides fresh insight into how the business actually operates. The project team should now take time to look at these findings, identifying feasible solutions in order to determine if performance improvements, as stated in the business plan, are achievable. A starting point for this work is to plot and analyse:

- *Material flows* – parts moving around the factory between the various operations. *This often shows that components are transported considerable distances between workstations, spending long periods waiting for machines or processes to become available.*
- *Information flows* – documents and work instructions required to release materials, process components, assemble, test and ship products. *This often shows that volumes of unnecessary paperwork usually add cost and little value to the process.*
- *Working practices* – present job structures and the restrictive practices that currently exist in the factory. *Highlights activities that would become more effective through people accepting responsibility for a variety of tasks, as opposed to the traditional approach of people focusing upon specific tasks.*
- *Manufacturing technology* – manufacturing and assembly methods. *Identifies opportunity for investment in alternative production technology that combines operations, improves process capability, satisfies capacity requirements and reduces overall manufacturing costs.*
- *Information technology and systems* – current manufacturing control systems and information technology support. *Identifies opportunities to manage manufacturing information electronically, removing paperwork from the factory, providing analysis tools for converting data into information, maintaining records, collating information and integrating business systems.*

The savings and operational performance improvements as presented in the business plan should be justified based upon specific opportunities with the potential of delivering the enhanced performance. The project team must be able to identify, quantify and confirm that savings are achievable within the planned time scales, but also challenged with accepting more demanding targets as a result of this initial review and the management commitment to support a supply-chain and manufacturing systems design programme.

The project team having completed this initial data collection phase and gained a full understanding of the programme requirements must also compile a formal project plan for the overall programme. The task of determining the work content for the various work packages is difficult and normally imprecise but it is important to split the project into a number of inter-related work packages, scope the magnitude of the task and estimate the resources needed to complete the work. Formal project management methods and documentation must be applied to these type of projects, because the work content always increases and timely management decisions may have to be taken based upon available information. The project owner's primary responsibility therefore is to direct the overall work programme, maintaining focus upon those areas considered critical to delivering improved operational performance within the time scales agreed in the business plan.

This initial phase should culminate with a formal management project review attended by the managing director, site general manager and other senior managers with the authority to sanction the project. They must agree the programme objectives and financial targets, releasing the funds needed to complete the manufacturing design work packages and provide full senior management support required to implement any proposed changes within the factory and supply-chain processes. The review also signals the start of the design project. Having gained full management commitment the additional resources needed to complete the numerous work packages identified in the project plans must now be made available.

At this stage or just prior to the formal management project review, some companies undertake an official project launch by taking members of the project team away for few days, providing training in the fundamentals of supply-chain processes and manufacturing systems design. This ensures team members have a common understanding of the concepts needed to commence the work packages, confirming the scope of work and allowing team members to take ownership of the tasks. These events also provide a valuable opportunity for team building and for people to gain an understanding of their contribution to the programme.

Chapter 2

Manufacturing system – module/cell identification

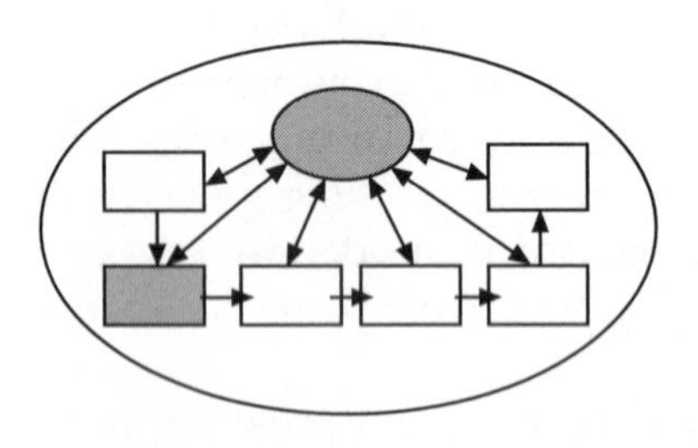

Manufacturing analysis

The second stage of the design process is to collate the critical aspects of the company business plan, market requirements, product and manufacturing information (gathered in the data collection phase) and develop a module and cell structure for the supply-chain and internal manufacturing systems. A process for defining the manufacturing module structure is shown in Fig. 2.1.

Modular manufacturing is fundamental to lean production and has proved *cost effective* in different market sectors, even those requiring high product variety and fluctuating production volumes. It will be used as the basis for the supply-chain and manufacturing systems design described in the following chapters.

Defining the manufacturing modules and cells is one of the most critical aspects of the manufacturing system design process, as it determines the structure of operations, deployment of resources, utilization of facilities and level of vertical integration. These factors are all decided early on in

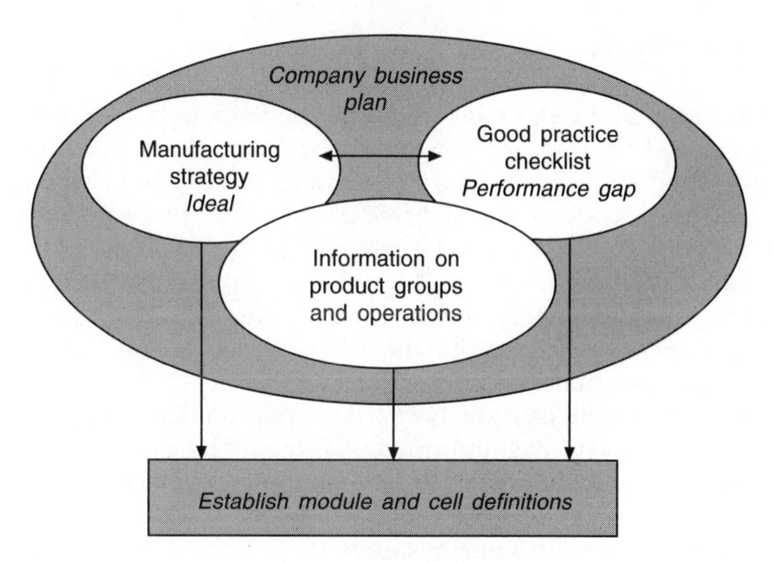

Figure 2.1 Module and cell definition process

the process, and once established are difficult to reverse. The term *'module'* is used to identify a number of *'cells'* required to manufacture a family of products when tasks are split between more than one natural group to simplify the process. The task of module and cell definition is imprecise; it relies on the experience, knowledge and vision of the team responsible for making the proposal, driven by the management team's perception of how innovative the changes must be in order for the business to become competitive. Managers facing 'significant emotional events' tend to have greater awareness of the need to make radical changes and are generally more committed to 'making things happen'.

The module and cell definition process is accomplished in three stages, but it is usually the relationships between a few key factors that fashions an optimum solution for a particular business. Each operates under specific conditions with a different customer base, ultimately leading to unique solutions. However, the process for designing the supply-chain and manufacturing systems can be based upon a common approach using similar tools and techniques. This process usually requires several iterations to identify the optimum structure, involving people at all levels within the organization to ensure a workable robust system.

The first stage is to identify natural groups or product families embodied within the range of products and components. The method for selecting groups is dependent upon the type of business operation and cannot be prescriptive, as each company has specific attributes fundamental to how it manufactures products to meet its customers' requirements.

Pareto analysis

The most powerful technique for identifying families is Pareto analysis; it can be used for deciding the important factors to be given greatest priority. The Pareto principle is linked to an 80/20 law believed in by Joseph M. Juran who considered that 80 per cent of the problems originated from 20 per cent of the causes. This 80/20 rule can be observed in many situations and provides a simple tool for establishing essential core elements. Pareto bar charts are drawn in sequential order representing the frequency of events. Plotting the data cumulatively (◆ – see Fig. 2.2) provides additional useful information.

Information accumulated in the first data collection stage is interrogated using Pareto analysis. Well-informed assessments help to identify family relationships needed to define the core modules and cells. Interpreting the results requires experience and fine judgement because factors have different weightings. In some instances the split may be nearer 60/40.

Modules and cells are formed by assembling all the resources needed to manufacture a family of products, a complete range of components or a collection of production processes.

They are characterized by:

- Clearly defined physical boundaries.
- They are equipped with the necessary manufacturing, assembly and test facilities needed to complete operations.

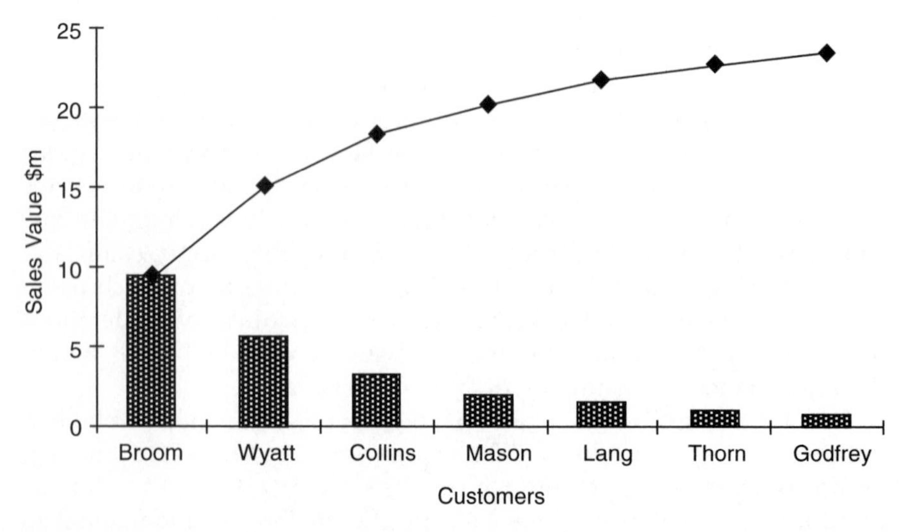

Figure 2.2 Typical Pareto of sales by key customer

Chart 2.1 Identifying natural groups

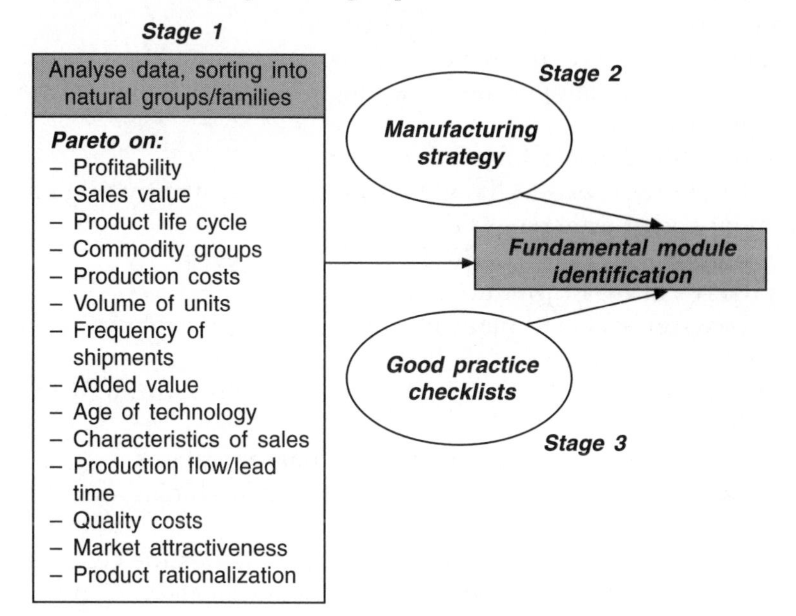

- Defined input and output gates for materials entering/leaving the area.
- Simple material flow patterns based upon short lead times.
- Obvious synergies between the range of products/components.
- Team of people taking responsibility for operating and owning facilities.
- Clearly defined scope of activities and understood by team members.
- Span of activities limited to a manageable size, allowing good team relationships and flexible working.
- Team members accepting responsibility for quality of own work.
- Team committed to achieving agreed output targets.

The project team responsible for the manufacturing system design must propose several viable solutions. These need refinement through open discussion and comparison with good practice adopted by other companies, in order to identify the most appropriate modular structure for the business. This may take several iterations to achieve an optimum cell definition; experience shows that changes at this concept stage are far easier to make than in the later design and implementation phases.

Features that may be evaluated using Pareto analysis to develop cell definitions include:

- Product types – identifying the main product lines that comprise the major sales volumes.

- Sales volumes by product group.
- Volume of products within a product group.
- Customers – examining the volume of sales and range of products.
- Variety of product families and the range of products within the family groups.
- Frequency of orders and fluctuations in demand.
- Value of the products and the gross margin generated by original equipment and aftermarket sales.
- Age of products and position in the product life cycle.
- Core process for major product lines.
- Technology – reviewing its maturity and potential for exploitation in new products or markets.
- Manufacturing routes – examining the work flow through the production facilities.
- Production processes and common machining operations.
- Value added stages in the process – test, assembly, machining, material processing.
- Competitiveness in performing tasks compared to outside vendors.
- Critical characteristics of a product or process that create competitive advantage.
- Flow of materials based upon runner, repeater, stranger classifications.
- Rationalization of product lines and consolidation of facilities.
- Capacity on bottleneck processes or major investments in capital plant.

An example of a *Pareto chart* showing sales values by customer and accumulated sales values is shown in Fig. 2.2. Similar charts should be constructed for the relevant parameters to identify natural groups. The primary requirement is to quantify the key factors that govern how the supply chain must operate to meet commitments made in the company business plan, sustaining long-term profitable growth for stakeholders in the company. Precise detail is far less important than establishing relative values and understanding the impact different factors have on operational performance.

> *In the final analysis, it is the management team's responsibility to confirm which features they regard as crucial to achieving a world class business, and to decide the overall module/cell structure.*

Manufacturing strategy – example document

The second stage of the module and cell definition process is for the operations management team to prepare a written manufacturing strategy.

One useful technique is to *'describe the factory you would wish to see when entering the front door'*. An example of a manufacturing strategy document for a systems component supplier is shown below (pp. 47–52).

Vertical integration and strategic sourcing

The business will carry out a strategic 'Make vs. Buy' analysis to identify those items that sustain the business's proprietary know-how, or enhance the customer's perceived value of the company's technical contribution. Consequently the major thrust of future investments will be directed towards the enhancement and wider exploitation of core technological and operational competencies.

Manufacturing activities will be modular, focusing upon design, assembly, test and customer support, manufacturing only those components that protect proprietary knowledge. The aim will be to provide maximum value added and retain confidential know-how in the marketplace.

Subassemblies and components will be manufactured only where necessary for competitive advantage and where special skill, know-how, processes and confidential competencies are employed.

All components, first stage operations and non-strategic subassemblies will be bought out and purchased based on minimizing total acquisition costs:

- Procurement will be from a number of selected, quality approved suppliers who are willing to act in partnership and share a proportion of the risks, including short-term fluctuations in material demand.
- Suppliers will provide long-term supply assurances protecting the company's position throughout the product life cycle.

A Supplies Module will be established to ensure the availability of quality approved materials to meet customer's schedules and manage supplier development. It will also provide the focus for supplier rationalization and implementation of make/buy strategies.

Location of factories

The company will retain the minimum number of factories needed to meet the business requirements and maintain an adequate return on investment. Their size, scope of activities and location will reflect customer requirements, access to specific markets and the company's need for cost-effective manufacturing to support a global business approach. Capacity requirements will also be considered on a global basis.

Manufacturing systems and equipment

When visiting a factory, the following attributes will be clearly visible:

- Excellent material organization and appearance of the facilities will provide a competitive edge to winning new business from existing and potential customers. Minimum space will be used for office support functions; these will be integrated with supply-chain and manufacturing operations in the factory.
- The manufacturing system will be designed to reduce lead times across the supply chain to secure competitive advantage from superior delivery performance that betters the market and customer expectation.
- All work will be team based, with cells organized around natural business processes with identified owners who champion continual improvement programmes and elimination of waste.
- Work groups and facilities will be designed to manufacture effectively a variety of products and product families with minimal changeover times.
- Cells will be identified using clear boundaries and appropriate colour schemes.
- Business systems will be designed to integrate with the manufacturing system for rapid response to changes in market demand and orientated to support manufacturing processes.
- All work groups will maintain prominent displays in the workplace communicating the key performance information and other issues relevant to particular modules.
- Technological process and equipment know-how will be developed internally to protect the company's competitive advantage. This will be approached systematically, using statistical methods to verify the capability of all key processes and measuring systems. It will be supported by concisely documented operational procedures and effective training. The goal will be to achieve zero defect production without any indirect inspection or rework.
- Full recognition will be given to the impact of existing and new processes on health, safety and the environment, demonstrating a commitment to be an environmentally responsible company.
- All cells will strive to operate to *Best Practice* which includes:

  - cross-training of staff;
  - continual development of effective standard working practices;
  - on-line quality systems supported by daily stand-up meetings;
  - progressive reduction of changeover times;
  - reducing lead times across the supply chain;
  - matching batch sizes to customer call-off quantities;

  - maintaining process capability for all processes and measuring systems;
  - total commitment to the elimination of waste through the application of recommendations for continual improvement;
  - bottleneck operations being identified and given priority;
  - total productive maintenance being adopted as normal working practice;
  - customer measures of performance and quality being used to focus improvement;
  - visible measures of performance to demonstrate achievements; and
  - self-audit procedures conducted as routine and all corrective actions implemented.

- Whenever possible, productivity will be maximized, all investment financially justified, and continuous improvement integrated with every-day working practices.
- Material movement and transfer between cells and workstations will be performed manually unless automation can be fully justified.
- Specific containers will be designed for the components: clean, stackable with the identity of the contents clearly visible.
- Machines should be fitted with stop-on-fault mechanisms, error prevention devices with consideration given to operation ergonomics.

Assembly and test activities will be integrated into *modules* that:

- Handle a range of products within a product family.
- House processes and equipment with sufficient flexibility to accommodate all predictable changes in product mix and production volumes.
- Are driven by a master schedule that accurately translates the customer requirements and ensures all components are available prior to starting the assembly process.
- Ensure products are assembled and tested by a team of multi-skilled workers capable of performing all the tasks, including final conformance tests, rectification, verification and packaging ready for dispatch to the customer.
- Pack products ready for use by the customer, in protective containers identifying the company.

Machining processes will be organized into cells designed to produce a range of critical components associated with a family of products in a way that does not compromise the manufacture efficiency of high volume and 'running' parts:

- Machining cells will undertake all the processes needed to complete the component, including deburring, part-marking, inserting, studding and building subassemblies, taking full responsibility for supplying quality assured parts directly into the assembly module.
- Production lead times will be reduced by minimizing the number of machining operations required to complete components, and attack the time components spend waiting for machining operations.
- Tooling will be dedicated to a family of components with the need for variety continually challenged.
- A team of people will take responsibility for operating a group of machines, including first-line maintenance, statistical process control and in-cycle operations.
- Material movements between major processes (e.g. component cells, heat treatment, assembly cells) will be synchronized to achieve continuous flow, minimizing work in progress and lead times.

Factory capacity

- Manufacturing modules will be staffed for a minimum of two full shifts with the target of achieving 135 hours per week as demand increases.
- Modern high capital investment facilities will be operated 24 hours per day, 7 days a week, using a low premium continental shift pattern.
- Contracts of employment will allow businesses to adjust manpower levels to meet economically the variations in customer demand, sometimes at short notice.

Production planning and control

The factory will work to an agreed *Master Production Schedule* established by cross-functional scheduling committees at three levels, owned by the general manager. This schedule will accurately translate commitments to the customer and will be underpinned by the production schedules, a production plan, cell schedules, capacity plans and supplier schedules, confirming material availability.

Manufacturing will work on a 'just-in-time' basis with customers and suppliers ensuring that inventory levels are kept to a minimum, while protecting customer shipments:

- The aim will be to respond rapidly to customer demand and to have products available for the customer before materials are paid for, thus operating on minimum working capital.
- Material flow within and between cells will be controlled by simplified

visible systems requiring reduced levels of central computer support and continual updating of data.

Measures of performance

The performance of every process will be visible and local module measures of performance will focus on:

- Customer satisfaction and the achievement of committed delivery dates.
- Achievement of the production plan.
- Reductions in work in progress and lead times.
- Quality performance and cost of quality.
- Cost reductions and achievement of target product costs.
- Continual improvement programmes and the elimination of waste.
- Capability of processes and measuring systems.

Organization

Manufacturing operations will operate with a minimum of four levels from general manager to production team members, and overheads will be subject to continual review for possible cost savings.

Leadership style will encourage:

- Team responsibility for each process and module-based activities.
- Waste elimination through teamwork and project planning.
- Customer satisfaction and quality being used as the driving parameters for continual improvement programmes.
- Commitment to providing a clean, safe and effective working environment.
- Regular manufacturing audits by senior managers to ensure compliance and implementation of corrective actions.

Full-time project managers dedicated to achieving project milestones, on time and within budget, will be appointed for significant change projects and product introduction programmes. Full-time task forces will be formed to implement the operational improvements and strategic challenges identified in the business plan.

Quality systems

- The quality system will conform to ISO 9000, and be designed and implemented for minimum bureaucracy and accredited by an authorized independent third party.

- Relevant process and quality documentation will be available to manufacturing teams.
- All customer problems will trigger a structured corrective action procedure. This will involve an appropriate team of experienced people responsible for ensuring the necessary corrective actions are identified and implemented.
- The quality system will incorporate an integrated manufacturing change control procedure linked to the engineering change control system.

Human resources

All employees will be:

- Able to share, whenever possible, a common environment including both conditions of employment and working arrangements.
- Able to participate in a continuous personal development programme linked to an effective system of performance appraisal and monitoring of training needs, so that all are trained to be proficient in a range of skills, allowing flexibility, job rotation and improved job satisfaction.
- Committed to continual improvement.
- Able to work in a team and contribute to enhancing team performance through group initiatives.
- Proud of their work, and taking full responsibility for quality and verifying conformance to specification.
- Dedicated to providing customer satisfaction and enhanced quality, while continually reducing waste and unnecessary costs.
- Willing to work flexible hours when needed to meet variations in workload.
- Rewarded in a fair and equitable manner.

New product requirements

The manufacturing team will integrate its engineers into effective Product Introduction programmes. All new components will be designed whenever possible to be within the capability envelope of existing manufacturing equipment, tooling and measuring systems. There will be a clear policy for economically controlling variety among components, tooling, fixtures and assemblies including standardized tool-sets for each machine tool carousel.

New product introduction programmes will exploit existing competencies and obtain new competencies necessary to assure future competitiveness in the marketplace. Additional manufacturing capabilities will be developed when a significant technical advantage can be identified and supported by a sound business case.

Good practice checklists

The third stage is to rate operational performance against good practice checklists that have been developed to identify current best practice within a particular business sector. The following checklists are derived from my experience in the automotive and aerospace component supply industries and provides an example of the factors that should be considered when assessing supply-chain and manufacturing operational performance.

Operational performance can be assessed by scoring factors in the checklist. These can be rated on two parameters using a scale of 1-5. The first assesses the importance of the item to the competitive position of the business (●) and the second to illustrate the current level of attainment (✚). This process should be used to show the relative importance of each item, focusing on the areas from which the greatest benefits will be derived.

Table 2.1 Scoring system for assessing the importance of different factors

Score	● *Importance for improving competitive position*	✚ *Current level of attainment*
1	not required	not started
2	minimal benefit	being considered
3	element of good practice	planned
4	must be addressed	being evaluated
5	critical item	fully implemented

Overview

The checklists for each section have been devised using my experience of supply-chain and manufacturing operations. It is not possible to compile a generic list because of the diversity of manufacturing operations, but many of the items should be relevant to most machinery-based manufacturing facilities. Identifying areas for improvement is not generally a major problem. The important task is to establish which items are the most critical to the business and will lead to significant improvements in performance if designed into the supply-chain and manufacturing processes.

Strategic sourcing

All businesses should conduct a strategic 'make versus buy' analysis to identify those items that encapsulate the business's proprietary knowledge, or enhance the customer's perceived value of the technical contribution.

Strategic sourcing checklist

Ref.	*Strategic sourcing operational practices*	*Scores*				
		1	*2*	*3*	*4*	*5*
1	Methodology required for taking make versus buy decisions		✚		●	
2	Make versus buy key aspect of product introduction process					✔
3	Worldwide sources for high expenditure items			✚		●
4	Vertical integration in line with industry sector norm	✚			●	
5	Core competencies for the business defined and agreed					✔
6	Supplies module responsible for supplier performance					
7	Budget for developing sourcing processes					
8	Resources for new product introduction teams					
9	Cost models to provide accurate comparisons					
10	Supplier rationalization					
11	Strategic sources for non-core components					
12	Supplier development resource to support key suppliers					
13	Resources for supplier quality assurance programmes					
14	Long-term contracts					
15	Parts classified on bills of material for purchased items					
16	Training material developed to increase skill base					
17	Personal development plans to expand purchasing skills					
18	Supplier quality measures					
19	Supplier delivery performance					
20	Supplier quality approval to deliver direct to assembly					
21	Internal manufacturing cells redesign after make versus buy					
22	Components analysed into runners, repeaters, strangers					
23	Supply-chain interfaces for consistent work flow					
24	Parts from suppliers arrive ready for use in the factory					
25	Parts protected from physical and environmental damage					
26	Simplified paperwork to remove unnecessary tasks					
27	Electronic links for paperless transactions					
28	Methods for agreeing production schedules					
29	Rules for setting and changing delivery schedules					
30	Suppliers support continuous improvement programmes					
31	Suppliers involved in product introduction process					
32	Suppliers acting in partnership and sharing risks					
33	Suppliers providing assignment stocks					
34	Suppliers able to provide long-term supply assurances					
35	Suppliers paid on time as agreed in contract					
36	Process capability of suppliers' equipment					
37	Key suppliers making daily shipments					
38	Strategic suppliers implement customer-focused modules					
39	Self-billing with major suppliers					
40	Blanket orders with preferred suppliers					
41	Low value standard items on automatic reordering system					

KEY ● importance for improving competitive position ✚ current level of attainment
✔ factor fully addressed

Factory space and location

Companies must retain a minimum number of factories needed to meet the business objectives and maintain an adequate return on investment. Many businesses have acquired excess factory space, duplicating internal resources for similar components.

Factory space and location checklist

Ref.	*Factory space and location operational practices*	*Scores*				
		1	*2*	*3*	*4*	*5*
1	Matrix of facilities information for other sites					
2	Site rationalization plan					
3	Consolidation of core component manufacture					
4	Worldwide locations for new facilities					
5	Overall company-wide site strategy					
6	Property database					
7	Impact on profit and cash of relocating sites					
8	Reaction of the customer base					
9	Site disposal					
10	Costs of environmental clean-up fully understood					
11	Impact upon supplier base					

Manufacturing system and equipment

These differ with the type of processes performed within the factory but the following attributes should be considered.

Manufacturing system and equipment checklist

Ref.	*Manufacturing system operational practices*	*Scores*				
		1	*2*	*3*	*4*	*5*
1	Manufacturing facilities appropriate to product groups					
2	Appearance of facilities					
3	Layout of factory orientated to customer requirements					
4	Minimum factory space devoted to production offices					
5	Support functions integrated into factory activities					
6	Supply chains designed to meet the business needs					
7	Manufacturing systems developed for minimum lead time					
8	Bottleneck process given priority and not interrupted					
9	Everybody committed to achieving delivery dates					
10	People working in small teams with identified team leader					
11	Team leader organizing work and identifying problems					
12	Regular group meetings to resolve problems					

(Contd.)

Ref.	*Strategic sourcing operational practices*	*Scores*				
		1	*2*	*3*	*4*	*5*
13	Continual improvement groups, supported with resources					
14	Teams structured to produce a defined group of parts					
15	Team taking responsibility for quality of own work					
16	People multi-skilled to perform variety of tasks					
17	Goal of zero defects with no indirect inspection accepted					
18	Teams flexibly structured					
19	Cells identified with clear boundaries					
20	Factory floor coated, kept clean and well maintained					
21	Working areas kept clean by team					
22	Team responsible for routine checks on equipment					
23	Management and team audits made on state of plant					
24	Tooling and fixtures cleaned and stored appropriately					
25	Preventive maintenance built into loading schedules					
26	Equipment cleaned, bolts tightened, minor faults rectified					
27	Planned coolant, filter and slurry tank maintenance					
28	Metal particle skimmers on grinding machines					
29	Guarding on equipment					
30	Noise from machines					
31	Fluid leaks on equipment					
32	Space around equipment free from oil seepage					
33	Process knowledge documented and regularly updated					
34	Protection of proprietary manufacturing processes					
35	Lead time reductions					
36	Reduced changeover times for processes					
37	High stocks/work in progress					
38	*Takt* times used to establish batch sizes					
39	Level of process capability on equipment					
40	Provision of safe working environment					
41	Processes operating in accordance with legislation					
42	Standard working practices documented and followed					
43	On-the-job training provided by team members					

Assembly and test activities

Assembly and test are usually regarded as processes that must be retained in-house and integrated into core modules and cells.

Assembly and test activities checklist

Ref.	*Assembly and test operational practices*	*Scores*				
		1	*2*	*3*	*4*	*5*
1	Assembly system designed to handle a range of products					
2	Assembly and test equipment suitable for product groups					
3	Equipment flexible and able to perform necessary tests					
4	Activities driven by master production schedule					
5	All parts available prior to starting assembly operation					
6	Parts routinely cleaned and examined prior to assembly					
7	Products assembled and tested by multi-skilled team					
8	Team performs any rectification and final conformance					
9	Products packed ready for use by the customer					
10	Containers protect the product from physical damage					
11	Packaging provides adequate environmental protection					
12	Products clearly marked, identifying the company					
13	Assembly tools and fixtures dedicated to product family					
14	Test equipment specified in control documentation					
15	Appropriate containers available for kits of parts					
16	Assembly area conforms to appropriate cleanliness level					
17	Practices documented emphasizing critical characteristics					
18	Equipment calibrated against known standards					
19	Verification system and calibration records maintained					
20	Routine maintenance built into test schedules					
21	Correlation checks made on similar test stands					
22	Test rig services: air, fluids, etc. checked to specification					
23	Power tools to aid assembly operation					
24	Ergonomically design for assembly stations					
25	Lighting levels					
26	Repetitive tasks automated to prevent operator strain					

Machining facilities

These should be organized into modules and cells, designed to produce, in the shortest possible lead time a range of critical components associated with a family of parts without compromising the efficient manufacture of high volume items.

Machining facilities checklist

Ref.	*Machining facilities operational practices*	*Scores*				
		1	*2*	*3*	*4*	*5*
1	Machining modules/cells for families of parts					
2	Cells performing all tasks needed to complete components					
3	Team responsibility for providing quality assured parts					

(Contd.)

Ref.	Machining facilities operational practices	Scores				
		1	2	3	4	5
4	Delivering components directly to the assembly module					
5	Systematic removal of machining operations					
6	Continual cost evaluation for alternative processes					
7	Lead times of components awaiting machining					
8	Machine running speeds					
9	Teams of people responsible for groups of machines					
10	*Nagare* principles for the machining systems					
11	Module/cell structures to maintain even work flow					
12	Different manufacturing system for runners and strangers					
13	Bottleneck machines having priority for work/maintenance					
14	Teams responsible for daily checks, first line maintenance					
15	Machine tools monitored for process capability					
16	Measuring systems evaluated for process capability					
17	Calibration of measuring equipment					
18	Records maintained on status of equipment					
19	Preventive maintenance on key machines					
20	Key machines checked annually against specification					
21	New plant purchased to specified process capability level					
22	Jigs and fixtures checked, and cleaned, prior to storing					
23	Dedicated tools controlled, preset and allocated to parts					
24	Programmes to reduce tool variety used in production					
25	Documented procedure for managing/controlling tools					
26	Components cleaned after machining operations					
27	Containers used for moving parts cleaned					
28	Containers able to be stacked without damaging contents					
29	Parts located separately preventing physical contact					
30	Material movement synchronized for even work flow					
31	Minimum materials held in cells					
32	Parts held in lowest cost state					
33	Manual methods of material transfer					
34	Mechanization stops process on fault detection					
35	Error prevention devices integral with process					
36	Temperature and humidity control to within specific limits					
37	Levels of contamination in cleaning tanks					
38	Appropriate lighting for flaw detection and inspection					
39	Overall lighting levels better than statutory requirements					
	Plating and heat treatment					
40	Regular checks and monitors on all plating solutions					
41	Checking condition and cleanliness of tanks					
42	Safe disposal and handling of waste products					
43	Plating thickness checked over the complete surface					
44	Processes verified using control chart, confirming capability					
45	Systematic checks on furnace temperature distribution					
46	Thermocouples/heating elements working and calibrated					
47	Test pieces compatible with part being treated					
48	Area clean with good extraction and ventilation					

Internal factory capacity

Determining and controlling manufacturing capacity is one of management's most difficult tasks and adjusting capacity to meet fluctuations in demand is generally achieved through changing work patterns or the number of people employed.

Internal factory capacity checklist

Ref.	*Internal factory capacity operational practices*	*Scores*				
		1	*2*	*3*	*4*	*5*
1	Method for determining capacity of the plant					
2	Bottleneck processes for pacing production					
3	Output constraints and bottlenecks					
4	Method of measuring output from the modules					
5	Key machines running three shifts					
6	Other areas working two full shifts					
7	Maintenance schedules included in overall capacity plan					
8	Methods for adapting capacity to customer needs					
9	Alternative shift patterns					
10	Total capacity across all sites for consolidation					
11	Contracts of employment and workforce flexibility					

Production planning and control

All supply chains need to work to an agreed master production schedule.

Production planning and control checklist

Ref.	*Production planning operational practices*	*Scores*				
		1	*2*	*3*	*4*	*5*
1	Master production schedule owned by general manager					
2	Master production schedule identifies long-term capacity					
3	Production schedule establishes materials requirements					
4	Production plan giving commitment for reporting period					
5	Schedules cover *all* the demands placed on production					
6	Production plan verified against known factory capacity					
7	Team determines own work lists to meet commitments					
8	Team leaders agreeing commitment to delivering plan					
9	System used for coordinating people, machines, materials					
10	Material categorized by usage, value, source, application					
11	Alternative scheduling methods used for different cells					
12	MRP simplified by removing low value items					
13	Special small items kitted by vendor ready for use					

(Contd.)

Ref.	*Production planning operational practices*	*Scores*				
		1	*2*	*3*	*4*	*5*
14	Accuracy on the computer database greater than 98 per cent					
15	Factory works *just-in-time* pulling work into assembly					
16	Computer up-time greater than 99 per cent					
17	*Kanban* used to pull work through cells, when applicable					
18	Material flow between cells controlled by simple systems					
19	Once in process, part is completed with minimum queues					

Measures of performance

The performance of the key processes should be displayed; team members are responsible for collecting the information.

Measures of performance checklist

Ref.	*Measures of performance operational practices*	*Scores*				
		1	*2*	*3*	*4*	*5*
	Customer measures					
1	Customer performance – *quality, price and delivery*					
2	Cost of quality as percentage of sales					
3	Cost of quality assurance activity					
4	Cost of internal failure rework, scrap, rectification					
5	Cost of external failure, returns, recalls, modifications					
6	Product conformance, number of concessions accepted					
7	Equipment returns/dispatches					
	Labour capacity					
8	Hours worked in period					
9	Number of units produced/hours worked					
10	Number of people employed, employees, subcontractors					
11	Sales per employee					
12	Added value per unit of pay					
13	Overtime worked					
14	Training hours per employee					
15	Unplanned down time					
16	Industrial accidents					
17	Environmental incidents					
18	Percentage cost bought out/sales					
19	Percentage cost direct labour/sales					
20	Percentage cost factory support/sales					
21	Percentage cost overheads/sales					
22	Number of key machine tools					
23	Percentage utilization of key machine tools					

Ref.	*Measures of performance operational practices*	*Scores*				
		1	*2*	*3*	*4*	*5*
24	Percentage of processes known to be process capable					
	Schedule adherence					
25	Value of planned MPS/sales					
26	Percentage achievement of the MPS					
27	Percentage achievement of OE deliveries to customer requirement date					
28	Percentage achievement of spares sent by customer request date					
29	Service level deliveries made to agreed turnround time					
30	Supplier delivery performance to purchase schedule					
31	Average lead time for in-house manufacture					
32	Average lead time for bought-out items/materials					
	Stocks					
33	Stock turns					
34	Percentage stock held as raw materials					
35	Percentage stock held as bought-out components					
36	Percentage work in progress					
37	Percentage finished goods					

Organization

Supply-chain processes should operate with a maximum of four levels, from general manager to shop floor, with overheads subject to constant review for possible cost savings.

Organization checklist

Ref.	*Operational practices*	*Scores*				
		1	*2*	*3*	*4*	*5*
1	Supply chain operates with maximum of four levels					
2	Overheads justified by determining contribution of added value					
3	Business organized around core processes					
4	One person owns the supply-chain process					
5	Core processes designed to meet business requirements					
6	Teams responsible for a product or process					
7	Teams responsible for all activities, led by team leader					
8	Continuous improvement integrated into normal working					
9	Clean, safe, effective workplace					
10	Project management disciplines to support change					
11	Training structured for manufacturing engineers					

(Contd.)

Ref.	*Operational practices*	*Scores*				
		1	*2*	*3*	*4*	*5*
12	Payment system recognizes skill levels and contribution					
13	Reward structure for professional engineers					
14	Engineers involved in customer development process					
15	Engineers seconded to product introduction teams					
16	Cell team members consulted on product introduction					
17	Resources and time allocated for improvement projects					
18	Time allowed for training					

Quality systems

All businesses should develop a quality system conforming to ISO 9000. It must be designed and implemented to require minimum bureaucracy and certified by an accredited third party.

Quality systems checklist

Ref.	*Quality systems operational practices*	*Scores*				
		1	*2*	*3*	*4*	*5*
1	Business quality system conforming to ISO 9000					
2	Quality system written for a minimum of bureaucracy					
3	Quality system verified against actual working practice					
4	People trained and tested in the procedures					
5	Accredited by approved third party					
6	Customer problem triggering root cause analysis of events					
7	Team approach to implement corrective actions					
8	Record maintained of changes to processes and procedures					
9	Key characteristics for monitoring controlled processes					
10	Understandable and appropriate documentation					
11	Documentation checking for conforming to latest release					
12	Methods verify compliance for critical characteristics					
13	Process planning sheets giving graphical work instructions					
14	Quality records on separate sheet to process instructions					
15	First article inspection report verified to work instructions					
16	Assembly processes supported by graphical information					
17	System-generated records subject to document control					
18	Records on calibration of measuring systems					
19	Recalibration performed at specified intervals					
20	Statistical methods for confirming machine tool capability					
21	Measuring systems verified for suitability of process					
22	Measuring equipment owned by the company					
23	Measuring equipment clean and protected when not in use					
24	Product conformance verified using customer specified techniques					
25	Conducting routine management quality audits					

Human resources

All employees should be able to share a common employment environment, including working conditions and arrangements.

Human resources checklist

Ref.	*Human resources operational practices*	*Scores*				
		1	*2*	*3*	*4*	*5*
1	Formal and informal communication systems					
2	Business goals disseminated, understood by workforce					
3	Challenges facing the business openly communicated					
4	Management listens to feedback from workforce					
5	Formal employee surveys to gauge perceptions					
6	Senior managers walk the factory					
7	Managers discuss issues first hand with the workforce					
8	Same conditions of employment and work arrangements for everyone					
9	Everyone able to participate in personal development					
10	Performance appraisals to monitor training needs					
11	Proficiency encouraged in a range of skills					
12	Job rotation and flexible working practices					
13	Importance of customer satisfaction being understood					
14	Total commitment, team work and group support					
15	Continuous improvement to enhance working practices					
16	Everyone able to stop processes compromising quality					
17	Teams dedicated to delivering quality while reducing costs					
18	Equitable reward systems, with fair return for effort					
19	Regular group meetings to inform and resolve problems					
20	People trained in problem-solving techniques:					
21	Pareto analysis					
22	Brainstorming					
23	Cause and effect diagrams					
24	Drawing graphs and Isoplots					
25	Process flow charts					
26	Process charting with respect to quantity or time					
27	Process capability calculations					
28	First-line maintenance requirements					
29	Records of people's skill range/levels of proficiency maintained					
30	Safety glasses worn in designated areas					
31	Hearing checks: routine and documented					
32	Eyesight checks routine					
33	Eating and drinking, only in designated areas					
34	On-the-job training; and people tested for proficiency					
35	People trained prior to perform the task					
36	Acknowledging alcohol or drug abuse and offering help					
37	Ethics programmes to ensure contract compliance					
38	Documenting health, safety and environment procedures					
39	Priority and immediate resolution of any HS&E issues					
40	Formal group reviews and actions on HS&E matters					
41	Availability of clean overalls giving pride of association					

Product introduction requirements

All businesses should integrate manufacturing engineers into the product introduction teams, ensuring new components are designed whenever possible, to fit within the capacity envelopes of existing manufacturing equipment and tooling.

Product introduction checklist

Ref.	*Product introduction operational practices*	*Scores*				
		1	*2*	*3*	*4*	*5*
1	Manufacturing fully involved in product introduction					
2	Products designed for manufacture					
3	Products designed for assembly					
4	Products designed for maintenance					
5	Manufacturing system designed for new products					
6	Existing facilities redesigned to reduce costs					
7	Investment to reduce product costs					
8	Evaluating alternative viable manufacturing methods					
9	Components designed for available production processes					
10	Suppliers involved in the design process					
11	Standard parts established and used whenever feasible					
12	Standard tooling used whenever possible					
13	Tooling for new designs fitting into current carousel slots					
14	Tool changes in carousels minimal for rapid changeovers					
15	Components designed for minimizing machine setup time					
16	Once inside, all operations to complete the part in-house					
17	Bill of materials owned and updated by design authority					
18	Record of engineering change maintained and controlled					
19	Fault recording and corrective action system established					
20	Product introduction design tools applied to process					
21	Process FMEA used to determine robustness of methods					

Each checklist item should have been scored; any factors important for improving competitive position, (●) rated 4 or 5 with low score for current performance, (✚) must be investigated, and taken into account when designing the supply-chain and manufacturing systems.

Module identification for cellular structures

Following these three stages:

- pareto analysis of the critical features governing profitability;

- development of a manufacturing strategy describing a vision of the factory; and
- evaluation of best practice checklists;

the optimum module and cell structure in many instances becomes self-evident. They may require minor modifications to accommodate existing capital investments, traditional skills or practical constraints within the business, but these differences can be systematically evaluated allowing rational management decisions to be taken.

If a modular/cell structure is not obvious, consideration should be given to examining the benefits that might be obtained from adopting an alternative manufacturing system more suited to the volume and variety requirements of the business. I will not attempt to explain these systems, because they are either jobbing shops for one-off projects, or for high volume, very low variety products justifying dedicated automated processes.

The earlier in the manufacturing design process that a module structure can be formally agreed, the less work is involved collecting information and analysing the present situation. In some instances this saves considerable time, but it must be stressed that to reconsider the module/cell structure later in the process, to take account of additional factors, involves considerably more effort than at this formative stage.

The actual module structure must be based upon the business requirements, but there is a growing tendency for product manufacturers to focus upon assembly and test linked to the production of (say) four or five core components that protect the company's proprietary knowledge. Selection of core components must be strategic and not based solely upon the installed manufacturing facilities because it:

- Determines future investments in both product and process technology.
- Changes the skills requirements towards purchasing, logistics and supplier quality assurance.
- Creates production capacity and space in the factory.
- Effects the cost bases of the business.
- Reduces the level of vertical integration.

If outsourcing is used to provide a short-term cost reduction opportunity, it may result in the outplacement of critical items with dire long-term consequences, for example having to purchase core components from a major competitor.

Supply-chain and manufacturing system characteristics resulting from the module and cell identification process (Fig. 2.3):

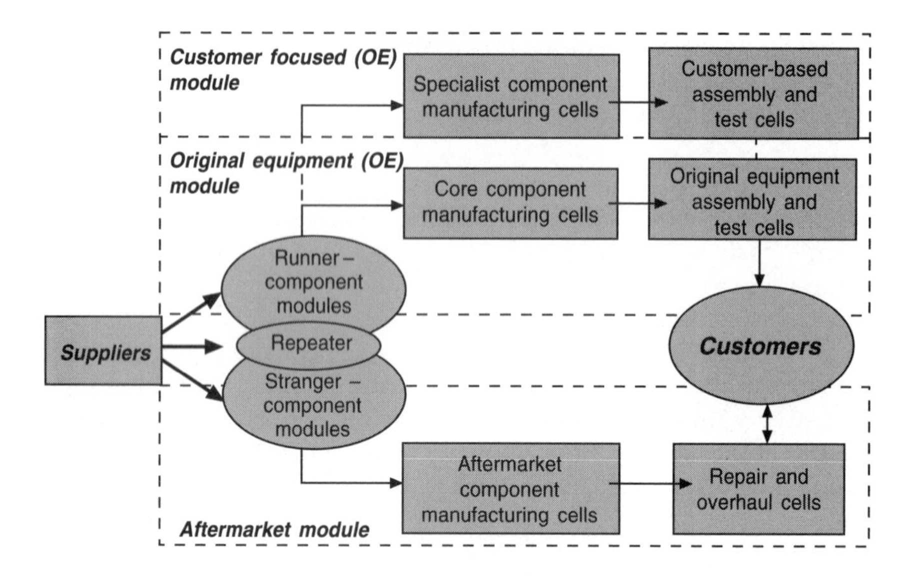

Figure 2.3 Example of module and cell structures

- Original equipment supply-chain and internal manufacturing system separated from aftermarket supply-chain processes.
- Separate modules may be established for specific key customers.
- Component manufacture for the two supply chains will be differentiated between runner, repeater and stranger parts. (This classification relates to the volume requirements, state of production engineering investment, tooling and ease of machine changeover.) Internal original equipment modules must have an even flow of work allowing production processes to develop a consistent rhythm that paces the factory.
- Stranger parts needed for aftermarket business that could disrupt the work flow and introduce an unnecessary burden of additional overheads will be manufactured in a dedicated module reflecting actual manufacturing costs.
- Original equipment manufacturing cells will be designed to produce a selected range of key components that protect the company's proprietary knowledge with the aim of having all the necessary processes to complete the component available within the cell.
- Component manufacturing cells, assembly area and test cells will be integrated into a single module with a clear identity.
- Non-core components will be purchased from preferred suppliers working in partnership and integrated into the supply-chain process.

Manufacturing module definition

Once the module/cell structure has been identified the next phase is to define in greater detail the actual core processes and technologies required within the cells, creating cell definitions (Chart 2.2).

The proposed module and cell structure needs further clarification to identify actual process technologies that will be employed within each cell. Therefore, concise manufacturing process specifications are required, identifying which technologies are needed to manufacture components or perform selected processing tasks. This procedure can be simplified if process descriptions are used to aggregate a number of operations with the range of technologies restricted to those considered core. Information is available from a number of sources and it is important to ensure that it is verified by examining manufacturing process layouts and conducting shop-floor audits. Once these core technologies have been identified they can be evaluated to assess the relative internal competitiveness of performing the task compared to an outside supplier.

Chart 2.2 Supply-chain and manufacturing systems definition process

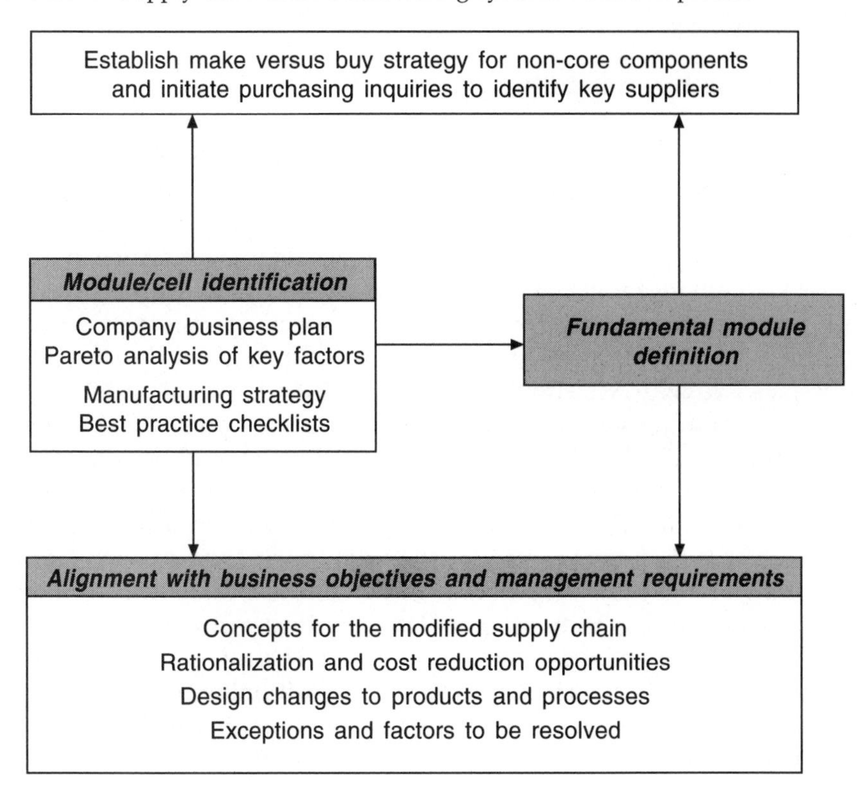

Strategic make versus buy analysis

Make versus buy decision making is one of the most important aspects of manufacturing system design. It impacts:

- Critical elements of the company structure.
- Employee skills requirements.
- Long-term product and process innovation.
- Investment in supply-chain processes and facilities.

Such decisions must be strategic and support the manufacturing objectives identified in the business plan. Outsourcing founded on short-term cost savings may result in losing control of key technologies crucial for long-term growth, or even survival. Therefore, they should be based upon quantitative information collected in the data collection phases. This is used to systematically translate the proposed module and cell structures into technical module definitions, identifying actual processes within each cell. The team must also focus upon the capital investment required to ensure that critical processes will have sufficient capacity and be fully process capable.

Make versus buy methodology

The first assessment must be of product and process technologies, taking judgements on the core processes to be retained in-house to protect proprietary product and related manufacturing knowledge critical to sustaining competitiveness in the marketplace. Manufacturing expertise provides considerable competitive advantage and, unlike product technology, can be protected through keeping specific details of the process confidential. To apply techniques used for formulating technology strategies two matrixes should be constructed (Charts 2.3 and 2.4).

The key product and processing technologies for a particular business sector are identified as 1, 2, 3, 4 and so on. These should be considered based upon the *competitive position* and *importance to the business* using the two differentiating factors (*needed to compete in particular market segments and available to the business*). Circles should be positioned on each matrix using the circle size to represent the sales volume expected from different product and process technologies. The length of arrow shows rate of change and future direction of a technology. The two matrixes should demonstrate considerable *similarities* in the positioning of technologies; any major discrepancies for important technologies require urgent management review.

Chart 2.3 Relative importance of different product and process technologies needed to compete in particular market sectors

Importance to the business
Low
Medium
High
Market competitive position
Strong
Neutral
Weak
1
2
3
4

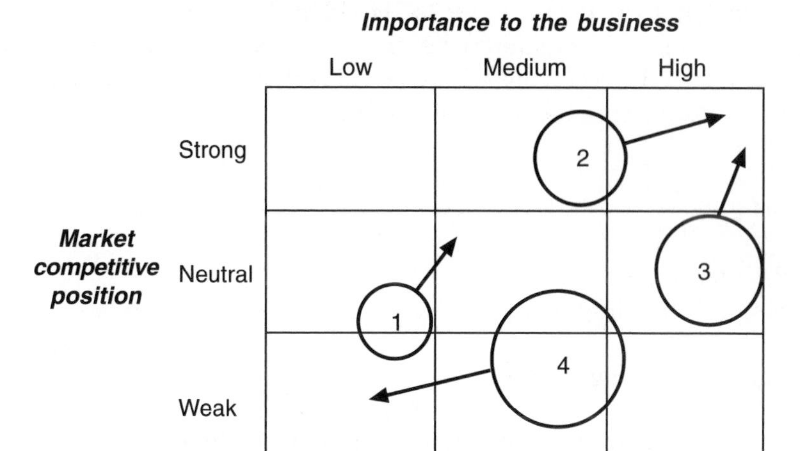

Chart 2.4 Relative importance of product and process technologies available to the business

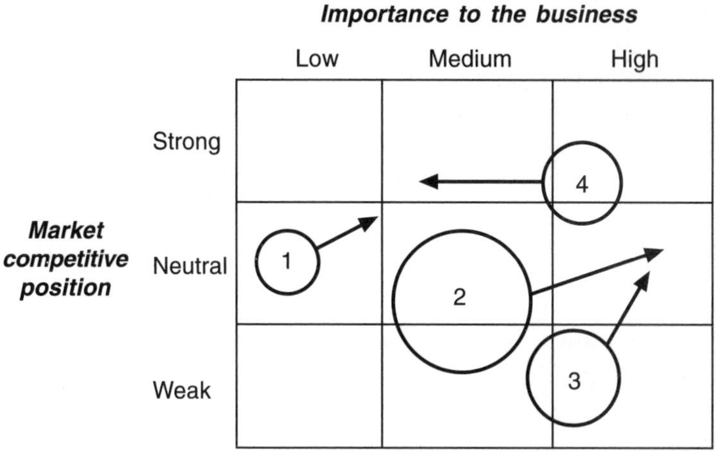

Importance to the business
Low
Medium
High
Market competitive position
Strong
Neutral
Weak
1
2
3
4

Technology cost models

Traditional product costing systems based upon applying 'blanket' factory overhead recovery rates do not provide sufficiently accurate product manufacturing costs. Therefore, to calculate the internal competitiveness of a process, a technology-based cost model is required. This should provide a more accurate assessment of actual costs, which can be used as a comparison with outside suppliers. This is achieved through allocating elements of factory costs and overheads in proportion to their actual usage, to determine specific costs per hour for each core technology. Using current management accounts and a breakdown of various cost elements, Pareto analysis can be used to identify major cost drivers. The significant cost factors should be classified and allocated to material costs, associated technology and general overhead charges, which can be used to determine more accurate product costs.

For example:

Material costs

- Raw materials.
- Bought-out components.

Allocated technology costs

- Labour costs for people employed operating the equipment.
- Depreciation charges for equipment.
- Specific surface treatments.
- Cost of tooling and fixtures.
- Space rental proportional to area occupied.
- Consumable materials.
- Maintenance charges.
- Scrap and rectification costs.
- Operating leases on equipment.
- Fuel costs.

Overhead charges

- Management supervision.
- Administration and support services.
- Head office charges.
- General facilities.
- Others.

The objective is to allocate the major elements of cost, without squandering resources analysing unnecessary detail. Focus must be maintained on the variable cost elements (labour, maintenance, tooling, etc.) because these items are directly impacted by the make versus buy decision. It must be remembered that in the short (and possibly long) term some variable charges will remain as fixed costs to the business, even if the technology is transferred to an outside supplier. Recovering these overheads must also be taken into account when comparing relative costs–increasing the need for lower costs from the supply base to be competitive, or for the business to rationalize these overhead costs. Ultimately, all costs are variable and the time element must also be considered when taking a strategic decision to outsource a technology.

Considerable judgement is needed to determine how to apportion costs; as a significant element, say 40 per cent, of directly allocated costs may still require assigning on a pro-rata basis. Also the total acquisition costs must include material cost variances, and internal expenses incurred administering orders, quality audits, supplier approvals, transportation, and so on. These need to be identified separately because they are *additional* costs associated with buying parts from a third party and added to supplier quotations as a percentage material surcharge supplement to reflect them.

A representative 'basket' of components should be selected for each technology, to establish the relative cost of internal manufacture compared to purchasing them from outside suppliers. The task is then to develop a factory-based technology cost rate for producing a particular number of units that can be used for direct comparison with quotations from third party suppliers. These two costs can then be used to establish an internal competitiveness index for the selected technologies; they can be placed on a supply-chain matrix showing the relative competitive position of the different production technologies used to manufacture products (Chart 2.5). The purpose of this chart is to confirm which key technologies add value and must be retained in-house, and which non-core activities cannot be performed competitively and should therefore be subcontracted. However, if a key technology critical to maintaining competitive advantage can be performed more cost effectively by a supplier, serious consideration must be given to making the necessary capital investment in a modified in-house process.

The manufacturing process definition should be reviewed against the cost index for each technical process to determine how the cells could be modified to take advantage of the supplier base. This often shows that initial machining operations can be performed more economically by a lower overhead supplier, and therefore parts should be pre-machined/ heat treated and delivered 'just-in-time' ready for higher value added processes to be performed in the module. However, once a part is brought into a cell, it should be completed in the cell as rapidly as possible. Moving parts in and out of cells to perform subcontract operations disrupts the flow of work and increases the risk of physical damage. If particular components need significant amounts of subcontract work, then serious consideration should be given to purchasing them as manufactured bought-out parts. In general, processes higher up the supply chain are more likely to add the greatest value and are the ones to be retained in-house.

Chart 2.5 Typical internal competitiveness index

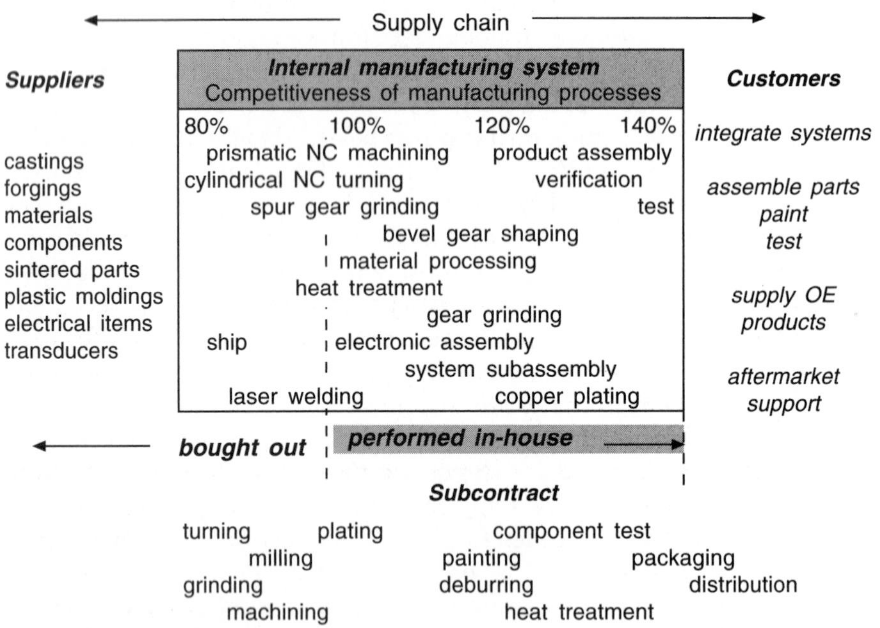

Framework for make versus buy decisions

Questions that should be asked to determine whether a particular technology is to be performed in-house or purchased from a supplier include:

- How competitive is the factory at performing the technology or process?
- Is the process linked to a business's strategic objectives?
- Is the process robust and capable?
- How important is the process to the business?
- Is the process a key technology for the company?
- How integrated is the process and will it impact other processes or products?
- What level of investment has been made in the process?
- How expensive is the process to maintain?
- What level of expenditure is needed to make the process competitive or capable?
- How critical is the process to achieving satisfactory business performance?
- Which new products require the process?
- How critical is the process to new product introduction?

- How environmentally friendly is the process and is there an alternative?
- How costly is waste disposal and treatments?
- Can the technology be purchased from a reliable supplier?
- How many suppliers could fulfil the requirements?
- Will particular supplier remain in business?
- Is the supplier able to deliver on time, production volumes needed to meet customer commitments?
- Is the supplier local to the factory?
- Has the supplier verified their processes are capable?
- What level of investment has been made by the supplier?
- Does it supply major competitors?
- Does the supplier's workforce have the necessary skills?
- Will your work be given adequate priority to ensure on-time delivery?
- Is the supplier financially secure?
- Can it purchase materials and services at competitive rates?
- Does the supplier's management team want the business?

One technique for assessing a make versus buy decision is to compile a matrix summarizing the critical factors for each technology, allowing comparisons to be made between different technologies. If outsourcing a particular one is feasible, then the benefits should be evaluated and quantified using the financial model to determine the level of savings. Outsourcing will probably influence manning levels, depreciation, work in progress, running costs and such; these savings must be reviewed to determine the impact on operational performance. The final decision, however, must be made regarding the effect outsourcing has upon profit, cash and return on investment for the business, using these improvements in business performance to challenge traditional manufacturing practice.

An evaluation as to whether a particular make versus buy policy is appropriate can be made by considering:

- What proportion of the sales revenue is spent with outside suppliers?
- What is the industry norm for major competitors?
- Does the present cost structure provide sufficient funds for investments?
- Are the facilities evenly loaded and adequately utilized?
- Is the product subject to seasonal demand?
- Is the level of fixed costs a problem for achieving a manageable break-even point?
- Does the manufacturing performance compare favourably with third party suppliers?
- Is the level of investment and age of equipment comparable with that of competitors?

A growing industrial trend is to translate fixed costs whenever possible into variable costs, making the business more resilient to fluctuating economic cycles. Adopting an aggressive make versus buy policy is a significant element in reflecting this trend.

Concept specification and management review

The final requirement for this stage of the module and cell definition process is to write a 'high level' manufacturing statement that complements the manufacturing strategy, making a firm recommendation for the modules and cell structure needed to create a world class business. The report should summarize the main elements of the supply-chain and manufacturing systems to be used as the basis of the detailed process design, making qualified statements on the expected level of performance following the implementation of the proposed changes.

For example:

Internal manufacture

- The business will maximize the ratio of added value to added cost, with internal manufacturing focusing upon:
  - items that protect the company's technical knowledge;
  - manufacturing processes involving core competencies;
  - in-house manufacturing processes that increase customer perceived value;
  - simplifying the flow of work where third party involvement would complicate the supply chain;
  - installed in-house capacity giving market protection and a barrier to entry; and
  - technical knowledge of the process providing competitive advantage.
- Non-strategic components, subassemblies and initial machining operations will be purchased from quality assured, preferred suppliers.
- Subcontract components will be procured as bought-out items whenever feasible

Manufacturing systems

- All work will be team based and organized around modular and cellular groups with identified leaders responsible for:

- maximizing productivity of the machines to run *(80 per cent)* of available hours;
- reducing factory lead times to *(say, 5 days)*;
- achieve promised delivery commitments *(100 per cent)* on time;
- reducing the work in progress to *(5 per cent)* of sales value;
- minimizing set-up times to under *(10 minutes)* for all machines;
- eliminating all types of waste through continual improvement, increasing productivity by *(say, 40 per cent)*;
- progressively improving quality, eliminating defects to achieve quality rates better than *(100 parts per million)*, and
- reducing batch sizes to *(say, 20)* manufacturing in quantities requested daily by the customer.

- Assembly and test activities will be performed in-house for the following products/customers:

  ⇒ ——————————
  ⇒ ——————————

- The manufacturing cells will focus upon:

  ⇒ ——————————- etc.

(A detailed schematic diagram should be produced that shows the layout of the modules and associated cells with a description of the main activities, range of components and machining operations to be undertaken within each area. A general work flow diagram should also be prepared to confirm that the production process will be simplified and distances travelled by components dramatically reduced.)

- Process and equipment technology will be developed for the *following* key process that protects competitive advantage:

  ⇒ —————————— etc.

Factory capacity and equipment utilization

- Manufacturing modules will be operated for (*120 hours per week).*
- Bottleneck operations will be given priority and operated for (*144 hours per week).*

The following information should be prepared for each product or customer module:

- Module definition including the cells required to support assembly.
- Sales volumes.

- Sales by product lines.
- Life projections and expected changes in volume over planned period.
- Profit generated by the portfolio of products manufactured in the module.
- Profit by product line or customer line.
- Number of units required for each of the product lines.
- Sales of new products to be introduced into the module in the planning period.
- Change in product mix over the planning period.
- Target performance figures for the module and associated component cells.
  - sales per employee;
  - added value per employee;
  - added value per unit of pay;
  - percentage cost of bought-out items to sales value;
  - stock levels and stock to sales ratio;
  - manufacturing lead times;
  - shift patterns;
  - utilization of key equipment;
  - cost of quality;
  - delivery compliance; and
  - frequency of shipments.
- Manufacturing activities and facilities:
  - assembly capacity;
  - age and state of equipment;
  - definition of core manufacturing cells;
  - test facilities and level of resources:
    - ◊ equipment requirement and available capacity;
    - ◊ processes that should be adopted by revised manufacturing process;
    - ◊ state of facilities;
    - ◊ age of equipment; and
    - ◊ assessment of process capability.
- Assessment of skill levels and shortages.
- Evaluation of training requirements.
- Estimated revenue and capital investment required to implement the proposed manufacturing system.
- Cost of undertaking the manufacturing system redesign and implementation project.
- Financial statement on the return on investment.

- Statement on customer benefits.
- Statement on business benefits and reasons for investment.

This information must be rigorously challenged by all team members together with people responsible for operations in order to present a strong business case to senior managers as to why the team's recommendations and the overall project proposals must be supported. Managers responsible for establishing the business's objectives must also be directly involved, taking full account of the issues presented and agreed in the business plan. The module/cellular structure must be a management decision, based upon Pareto analysis to identify natural groups, the manufacturing policy, findings from the strategic make versus buy analysis and the operating constraints imposed upon the business. For example, available factory capacity, capital expenditure limits, number of employees, inherent skills, condition of existing equipment, proximity to a good supply base for materials and such.

The final module/cellular structure cannot be determined through analysis alone and the management team must accept full responsibility making this ultimate decision, together with confirming the future sales volumes on which to size manufacturing resources and facilities.

An appropriate technique for ensuring the necessary level of understanding and gaining support for the proposals is to hold regular management reviews and workshops throughout the project. These should be used to seek information and guidance on findings, including testing more ambitious strategies that could be adopted.

However, at this stage in the project, a formal approval procedure is required ensuring full agreement on the module and cellular structures to be used in the subsequent detailed design phase of the project. A formal project review should be held by the managing director, supported the group finance director, programme director and the local management team, who witness a full presentation, commenting on and amending (if necessary) the project team's recommendations. The local management team should then prepare and sign a summary paper confirming that they support the proposed changes and commit to achieving the agreed improved business performance. This must be countersigned by the managing director and financial director who agree to provide the financial support needed to implement the proposals.

This level of manufacturing redesign impacts the total business and must be included as a significant element of the company business plan. The local management team must be intimately involved throughout, providing direction, establishing priorities for overall business initiatives and seeking opportunities for obtaining quick business benefits. If possible, other significant change activities should be kept to a minimum in order to avoid management overload and total loss of control. The next phase of the manufacturing design involves re-evaluating current operating practices. Once implemented recovering from serious unplanned events using traditional methods may not be viable, therefore the new process requires considerable attention to detail, ensuring all reasonable eventualities have been asssessed.

Chapter 3

Steady state design of manufacturing modules

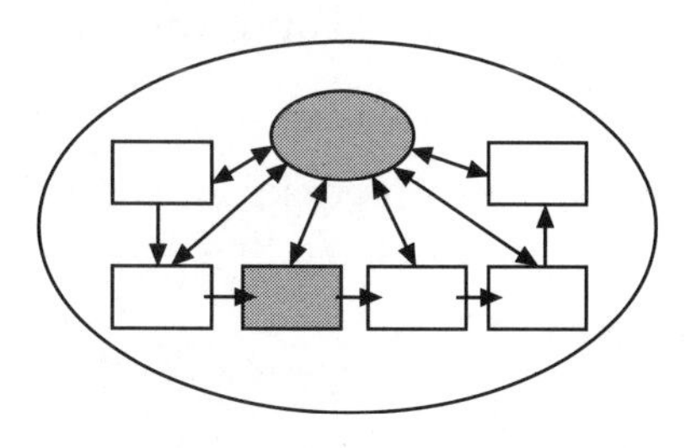

Characteristics of modular factories

The fundamental concept of modular factories is to bring resources together in natural groups, forming cells that are responsible for a product or process. Physical resources include plant, equipment, tooling, work instructions, planning systems and so on that are needed by a team of people to produce good quality products. Several cells are grouped together to form a product module. The term 'product' is applied quite widely and could be similar product lines, components, processes or assemblies, including specialist cells supporting more than one product module. Most factories require a combination of modules depending upon the volume and variety of product being manufactured, but the aim is to establish a number of discrete, self-managed modules designed to meet specific customer requirements and business objectives.

A cell requires the following features:

- Clearly marked physical boundaries.

- Defined gateways controlling work entering/leaving the cell.
- All equipment needed to complete jobs situated within the cell. (If a job has to leave a cell on more than two occasions, then the cell definition should be reviewed for a possible alternative route.)
- An identity that people relate to and are proud to be associated with.
- A family of similar products and/or processes.
- Cell team owning the processes, taking responsibility for delivering good quality work on time.
- The correct size to perform the task and create a team environment.
- A self-manageable size.
- Team members capable of performing a range of activities within the cell.
- Everyone understands who their customers are and their requirements.
- Key decisions on production control, capacity and quality taken by the team member most knowledgeable about the current situation.
- Cell team assigns priorities, allocating resources based upon the local circumstances and customer requirements.

Cells must be designed to reduce the number of process routes and simplify the flow of material through the supply chain leading to:

- Easier control of material movements.
- Predictable processing times through the cell.
- Lower levels of work in progress.
- Reduced lead times for parts to be manufactured.
- Increased work flow reducing overall factory lead times.
- Predictable demand forecasting through working with shorter lead times.
- Greater stability and pace increasing effectiveness and customer service.

The accountability for operational performance remains within the cell and performance monitors should be devised that encapsulate the business objectives (*delivery on time against customer schedules, quality performance, level of working capital, control of operating expenses*). These monitors must be used positively, demonstrating that people's efforts have been recognized by senior managers and that cellular structures are operating effectively.

Creating module and cell structures

The previous chapter on module/cell identification and definition explained the procedure for strategically determining the structure of the supply-chain and in-house manufacturing systems. It is now important to focus upon these requirements as opposed to further analysis of existing processes. The manufacturing statement formally agreed at the senior management review provides the overall definition for the module and cell structures. The next task is to complete a *detailed module/cell design*; this can be based upon a number of common factors such as: product types, customers, volumes, varieties of products, or manufacturing routes. Manufacturing processes can also be used: assembly, test, heat treatment and precision machining operations. Alternative solutions should have been evaluated in the previous phase using Pareto analysis to develop cell definitions. These preferred options now require designing in detail to define the actual production processes needed within the cells. The opportunity must be taken to radically evaluate different production methods. Reinstalling traditional processes would mean cells remained functionally orientated, with failure to exploit the benefits of cellular manufacturing. The objective is to produce a module structure providing a business with product-focused production facilities that are capable of being managed by teams of people working within the modules. The most effective structures tend to be simple based upon groups of products (Chart 3.1).

The initial design of modules/cells requires consideration of the aspects shown in Chart 3.2.

Chart 3.1 Example of simple module structure

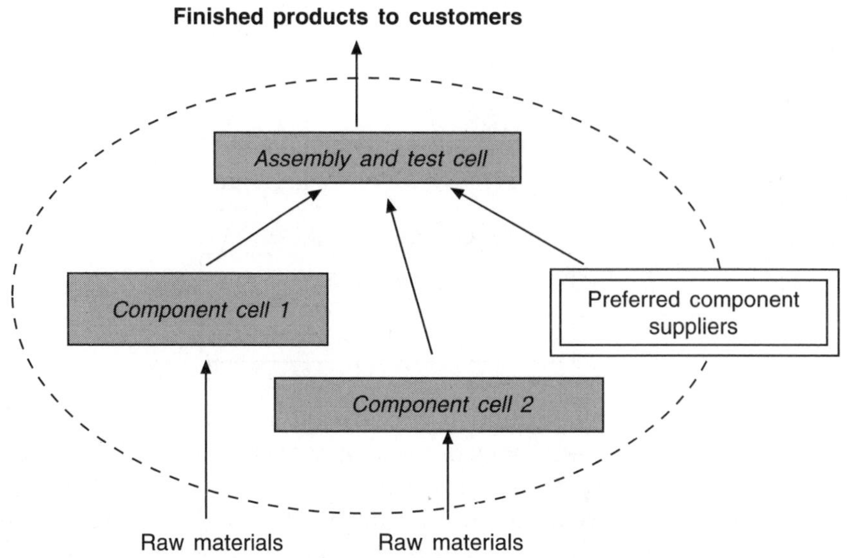

Chart 3.2 Factors to be evaluated in the steady state design

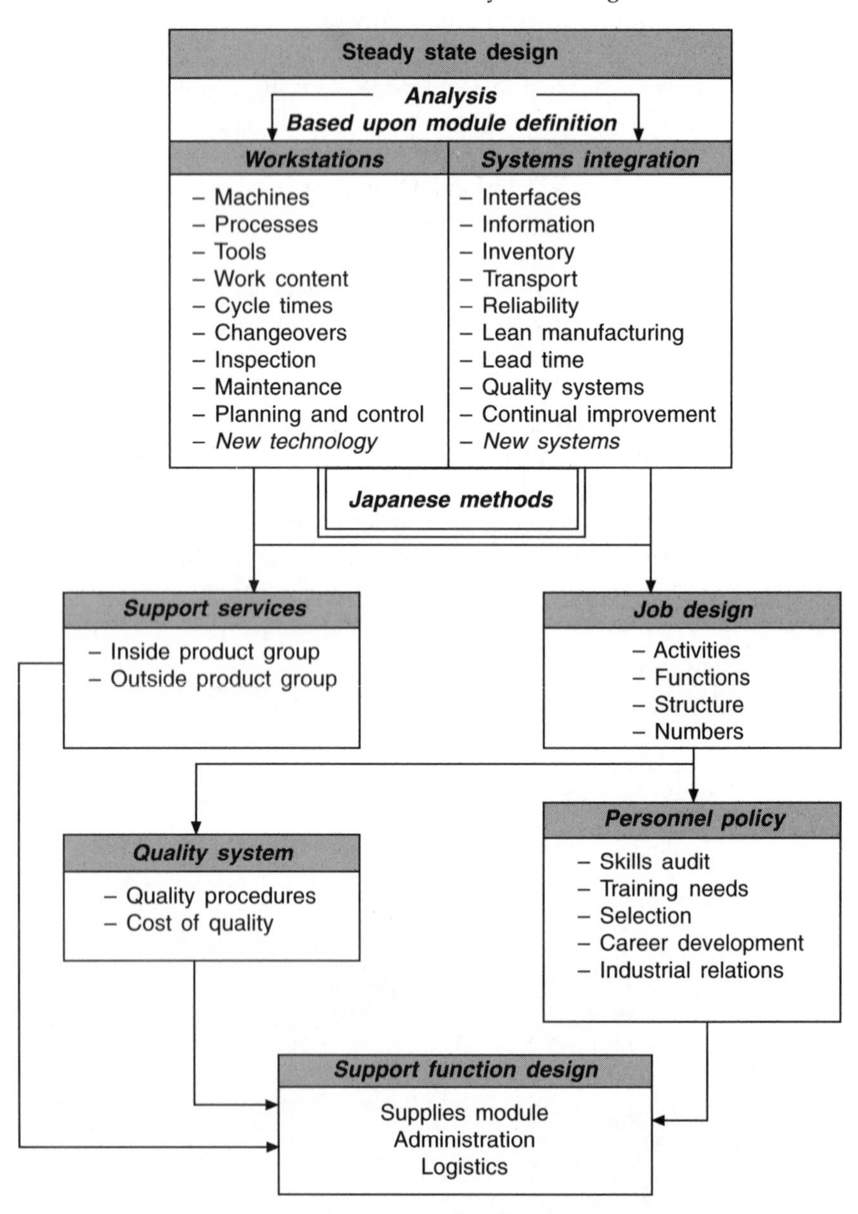

When designing the system it not always possible to make cells completely self-contained. For example, it may not be feasible to duplicate expensive equipment due to its low utilization or the particular nature of a process. This situation may be accommodated to a degree by subcontracting work between cells and outside suppliers, or establishing separate cells around a key process. In some instances process cells form a 'natural break' between different stages of manufacture, for example pre- and post-heat treatment. The system may also need a flexible non-focused cell to accommodate parts that will not fit into standard families. It would operate as a multi-purpose unit, suitably equipped to manufacture high variety in low volumes. The precise cellular structure should be established to meet the specific needs and circumstances of the business, but it is important to spend time considering possible alternative solutions until a near optimum structure has been reached.

Defining steady state

The steady state design of the cells is based upon events remaining constant with respect to time. *Average* figures are used for cycle times, changeover times, demand, volumes, product mix, machine breakdown, and so on. This approach simplifies calculations and allows the scope of the different parameters to be determined, but an assumption that factors *remain static* is oversimplistic and proposals must be dynamically analysed before any recommended solutions are accepted for implementation.

The purpose of steady state design is to:

- Identify and collate information on the bottleneck processes.
- Examine alternative systems and methods of working, identifying non-value added activities that could be eliminated.
- Compare new manufacturing concepts, examining the advantages and possible shortcomings.
- Evaluate alternative production technologies, investigating the benefits and return on capital investment.
- Confirm the proposed system has the capability to meet the ambitious performance criteria identified for the project.
  (*It is important to impose 'stretching' targets aimed at beating accepted best practice to ensure that people think beyond present methods. However, targets must also be attainable and realistic, as a relatively* modest *performance improvement may transform profitability.*)
- Scope the overall cells in terms of product families, required plant and equipment, manning levels, and so on. This is a relatively complex task, requiring considerable detailed information and analysis (see input/output Charts 3.3 (p. 100) and 3.4 (p. 103)). The first chart examines general requirements and the second detailed manufacturing information integrating cell operations.
- Confirm and quantify the resources needed for operating the cells.

This calculation is normally concerned with resources directly engaged in production activities:

- type of plant and equipment;
- number of machines;
- estimated cell loading and shift patterns;
- people required to operate the cell;
- material usage; and
- investment in tooling, fixtures and specialist processes.

- Identify bottleneck operations and analysis of events impacting performance.
- Evaluate the changeover times for critical process and methods for improvement.
- Refinements to the processes that remove non-value added activities and support continual improvements.
- Establish standard working practices.
- Introduce lean manufacturing principles that eliminate waste.
- Provide improvements to the flow of material, reducing overall lead times.
- Identify the indirect support activities.
- Establish training plans that identify skill requirements, training providers and associated costs.

Resource definition

The resources required by each cell have to be calculated using steady state conditions; in practice these never occur, but unless average values are used the variables make the situation too complex to analyse. Steady state values are assigned to:

- Range of products manufactured.
- Volumes of products required over the period.
- Machine and assembly cycle times.
- Operator performance levels.
- Equipment availability and process capability.

It is also assumed that:

- Operation times are correct.
- Process details conform to production layout.
- Scrap is constant.
- Unplanned events do not impact production.

This resource information is used to scope the detailed production process identifying:

- Type of processes and activities within each cell.
- Number of machines and equipment required.
- Number of people in team needed to operate processes.
- Effectiveness of alternative manufacturing methods.
- Materials and components to be processed at each workstation within the cell.
- Tooling requirements and consumable materials to support production.
- Levels of work in progress.
- Investment requirements in new plant and equipment and refurbishment of existing facilities.

The basic capacity requirement equation for a particular item is straightforward:

$$\text{Capacity required} = \text{Average demand rate} \times \text{Time spent processing each component}$$

The calculation is repeated for each operation and the results added together for all the products passing across a particular workstation. However, the available capacity calculation is far more difficult as differences occur between the capacity of the plant and that of the workforce. Operators who work more than one machine, unmanned running and movement of people across workstations add further complexity. The production capacity calculation must accommodate such flexibility accounting for its impact upon production. The results are expressed as factors of time and often referred to as *standard hours*. These figures, however, are still unrealistic for planning purposes because they do not include the time needed performing necessary non-productive tasks, such as personal time, thinking of possible process improvements, participating in team meetings. This additional time should be included as a utilization factor based upon previous experience.

Detailed performance factors for equipment could include time:

- spent changing machines to manufacture different products;
- required for making adjustments and quality assurance;
- planned for preventive maintenance;
- lost through unplanned events – tool failure, machine breakdowns;
- lost through processing inefficiencies against theoretical cycle times; and
- spent reworking scrap and rectification.

Non-productive tasks for team members include time spent:

- on indirect work – handling materials, changing machines over, first line maintenance, setting machine, cleaning area, deburring parts, masking components, and seeking tools/paperwork/materials;
- inspecting products and corrective actions;
- working on continual improvement programmes, waste elimination; and
- performing unmeasured work (no available standards).

Also, performance is affected by:

- personal interruptions and agendas;
- individual operator performance and group dynamics; and
- absenteeism and accidents.

It is important to establish a reliable method for calculating available capacities, because it will provide a foundation for determining the *'rough cut'* production capacity needed for production planning and control. This calculation varies between different facilities based upon the type of equipment and the way the equipment is to be operated. Each element affecting operational performance has to be considered to determine the limiting factors. For example, if machines run unattended then capacity is dependent upon machine loading and utilization, but if machines are multi-manned then a limiting factor may be the operators and changeover times. It is obvious that no general formula exists for making the calculation, as each situation has its own limitations and controlling circumstances. This level of complexity explains why manufacturing design projects must be followed by continual improvement programmes, addressing specific operational requirements that cannot be resolved by calculation or simulation.

Such calculations are suited to PC spreadsheet analysis packages. If the process elements are constructed systematically, the influence of the key factors on operational performance can be evaluated by changing various parameters. Time must be spent evaluating the impact of alternative solutions and determining a near optimum structure for groups of workstations within the cells. Several proprietary computer-based manufacturing simulation tools are available; these allow cell layout models to be constructed and simulated performance evaluated under a range of differing operating conditions. They are relatively easy to apply to most types of factory layouts, and if a number of modules have to be designed they can be a good investment.

The next stage is to develop similar spreadsheet models to determine the material requirement for each cell using the bill of materials and associated tooling information for the proposed operational layouts. The inventory and work in progress levels can be estimated using a typical mix of products and average volumes. These may need adjusting at the dynamic design stage, introducing some additional stock to protect the overall manufacturing system from unplanned events. The level of stock and work in progress is governed by a combination of practical considerations and stocking policy. However, the Japanese view of stock being 'evil' should be the guiding principle. Work must flow through the cell with minimal time spent waiting for processing and reduced equipment changeover times providing maximum flexibility.

Stock and work in progress levels are influenced by a number of factors:

- Customer call-off quantities.
- Length of time needed to changeover machines.
- Capability of process following changeover.
- Manning levels and methods used for changeover.
- Flexibility of equipment to produce smaller quantities.
- Type of manufacturing system.
- Layout of the production facilities.

However, the greatest influence on manufacturing effectiveness is the choice of manufacturing system, the layout of production facilities and the design effort applied to how the system will operate.

Investment in manufacturing technology

Investment levels in modern machine tools and test equipment depend upon the financial state of the business and capital expenditure negotiated in the business plan. Priority for capital investment must be given to ensuring that equipment and measuring systems are process capable, followed by removing bottleneck processes that govern overall plant capacity. Modern machine tools with powered tooling and state of the art numerical controls have the facility to combine machining operations and provide opportunities for transforming business profitability. The objective of capital investment in new equipment should be to load a component and perform as many operations as possible without manual intervention.

The level of investment committed to achieving this ideal must be quantified by:

- Cost justification for the expenditure in new equipment and tooling:

  - utilization of the equipment based upon the family of parts to be produced; and
  - savings on labour, maintenance, factory space, work in progress and production lead times.
- Improvements in customer satisfaction and quality:
  - reduction in scrap and rectification;
  - inspection and corrective actions; and
  - on-time delivery performance.
- Expected returns on investment and discounted cash flow requirements.

A general rule when acquiring new equipment is to purchase from a leading manufacturer standard machines with a known level of process capability to perform necessary operations within the available cycle time. The machines should be robust, reliable, straightforward to operate and proven in a similar production environment somewhere in the world. Careful consideration must be given to auxiliary items needed for particular applications, but it is important to introduce technical features that ensure full process capability. I have learnt by experience that installing oversophisticated integrated systems has failed to provide expected returns on investment, as linking several machines using automation (even with reliability levels of 99 per cent) usually results in unacceptable overall machine availability. The recommended approach is to develop an investment policy and identify those areas of the supply chain that will provide the greatest return on investment for the business. It is tempting occasionally to allocate all capital expenditure to modernize particular component cells. However, if this represents 20 per cent of the product cost with half being attributable to materials, a 50 per cent saving in labour and quality costs will only impact product costs by 5 per cent. If this capital were invested across the facility it may yield a much more attractive return.

In some instances the only way to achieve the necessary level of production and process capability is to invest in specialist machines, using emerging technology. These investments have the ability to transform the production method, providing real competitive advantage. In these circumstances it is important to allocate internal resources to develop the process, to prevent the machine supplier from gaining the full process knowledge needed to make the technology commercially viable to someone else.

Investment in assembly and test equipment cannot be neglected, this too must be fully process capable. The equipment required for assembly and test facilities is often bespoke and supplied by specialist manufacturers, but as a core competence investment in assembly and test equipment it has the potential to make a significant impact upon business performance.

Japanese manufacturing techniques

Japanese car manufacturers (particularly Toyota) have transformed traditional thinking on production methods. They introduced new concepts for manufacturing systems that have become the benchmark for manufacturing companies around the world. Such methods must be considered for the steady state system design and in many cases will provide a basis for establishing optimum layouts for equipment and corresponding working practices. The Toyota system shows considerable differences with conventional manufacturing practices but these have been proved without question to be more customer orientated and cost effective than traditional methods (Table 3.1).

Understanding the differences between these two systems gives a good appreciation of the Toyota production methods.

Designing Japanese style manufacturing systems

A manufacturing system design focuses upon six key elements; each one is analysed to determine the optimum solution for a particular cell with the objective of eliminating possible causes of variation and non-value added activities.

The elements are:

1. Identify and eliminate all possible waste in the system.
2. Design U-shaped lines to manufacture particular components or products.
3. Identify changeover requirements for equipment and production processes.
4. Design material handling and Kanban systems pulling components into assembly.
5. Evaluate Nagare principles for non-stock one-piece flow.
6. Establish standard operating procedures and train team members.

These items should be considered in depth at this manufacturing design phase, because it is these concepts that need embodying into the system. However, no matter how good the design, the system variables always need refining, through developing continuous improvement programmes once the system has been implemented.

Table 3.1 Conventional versus Japanese manufacturing practices

	Item	*Conventional system*	*Japanese system*
1	*Emphasis on industrial engineering*	Not a prime factor and dominated by man-hour and machine performance	Critical element that focuses upon employees, inventory and machine performance
2	*Assignment of key managers*	Achieve production goals using people	Achieve production goals while continually removing people
3	*Selling price*	Cost + profit = selling price	Selling price – profit = cost
4	*Production planning*	Draw up long-term, intermediate and daily schedules for production	Assembly process driven by customer and parts pulled through from cells and suppliers
5	*Layout*	Facilities arranged according to machine types, in straight lines	Facilities arranged to support the process, generally U-shaped
6	*Shift patterns*	Rigid; three shifts with three groups	Flexible two shifts with two groups
7	*Lot production*	Large lots, continuous production	Small lots, feeding one unit at a time to synchronize mixed production
8	*Lines*	Conveyors	Feeding one unit at a time providing synchronous flow
9	*Transport between processes*	Delivered	Fetched by team
10	*Work in progress and inventory*	Economical batches and computer-based systems	One cycle time, Kanban systems
11	*Number of changeovers and times*	Fewer changeovers as possible, taking a long time	More acceptable, time is continually reduced
12	*Trend in automation*	Fast, specialized, large installations	Slower speed synchronized with the cycle time, universal machines
13	*Number of machines per person*	One per person, division of labour and specialized operators	One person handling several machines and processes
14	*Working posture*	Sitting or standing	Walking
15	*Method of improvement*	Equipment and facilities	Continuous refinement of working methods, linked equipment improvement
16	*Waiting time*	Accepted as part of normal practice	Excluded or used for making improvement to process
17	*Overproduction*	Producing ahead of schedule is accepted as good practice	Making too much leads to waste

(Contd.)

Table 3.1 Contd.

	Item	*Conventional system*	*Japanese system*
18	*Quality control*	Statistical quality control or inspection	All processes are known to be capable and operated to produce consistently at the mean tolerance
19	*In-process checking*	Sampling or no inspection at all	Agreed sampling plan and continual verification by operators
20	*Operational improvement*	Identified and implemented by staff and managers	Identified and implemented by team members supported directly by managers
21	*Performance targets*	Measured by the day or week	Monitored by hourly targets
22	*Analytical methods*	Work flow analysis and other techniques	Timing operations in the work environment
23	*Establish standard times*	Average values for completing operation	Minimum time for combined tasks, the 'challenge time'
24	*Establish standard work instructions*	Identifying method is main objective	Sequence of events to complete the process is key

Elimination of waste

Toyota identified seven *wastes* in a manufacturing process that must be eliminated, and these form the foundation of a manufacturing philosophy:

1 Waste in manufacturing too many products.
2 Waste caused by waiting.
3 Waste in transporting materials.
4 Waste in processing.
5 Waste in inventory.
6 Wasteful movement.
7 Waste in making defective products.

These are addressed at the design stage by evaluating:

Waste in manufacturing too many products

This is manifest in two types of overproduction: making more than is required and the premature production of items needed in future. Overproduction and combining orders to create larger batch quantities may mask problems existing in the production system. These problems are generally associated with:

- The cause of defective products not being detected.
- Parts waiting for machines/people to carry out the work.
- Long changeover times for processes.
- Impact of machine breakdowns/components not being available from suppliers.
- Line imbalance/inadequate job design.
- Absenteeism.

Waste due to inventory
Traditional thinking considered inventories an asset, providing a cushion against unplanned events. The losses that arise from inventory include:

- Interest on the capital needed to fund it.
- Storage costs and the deterioration of products held in stores.
- Cost of the inventory control system and the people to run it.
- Transport costs and material handling equipment.
- Cost of space, shelving, pallets and packaging.
- Cost of obsolete stock.

It is estimated that (on average) a product costs a business 20 per cent of its value each year to hold as stock.

Basic concept of U-shaped lines
The concept focuses upon achieving a consistent *flow,* associated with three elements – *materials, information* and *manpower*. The objective is to produce good quality, just in time. A standard layout for assembly and machining processes usually takes the form of a U-shaped line, see Fig. 3.1 (p. 100), designed to operate under the following conditions:

- Cycle time is considered the main factor and the production flow time for each piece or operation is standardized.
- Non-productive procedures are designed out of the process.
- Equipment is installed in the sequence of the process layout.
- One operator performs several tasks, appropriate to the available cycle time, to complete the operations.
- Component and assembly cells are linked into a module to provide continual flow of finished products using Kanbans to trigger production.
- Cells have 'gateways' for parts entering/leaving the cell controlled by the same person.
- Machine layouts allow flexibility for team members to assist each other and continually balance the workload, maintaining production flow.

- Equipment layout accommodates increases/decreases in production volumes at any time.
- Opportunities to reduce manpower identified at the design stage, to take advantage of experience, learner curves and continual improvement.
- Materials moved in small quantities by cell team members using an appropriate materials handling system.

Changeover time reduction
Short changeover times are the foundation of success for high variety, low volume production systems; if they can be reduced to *zero* small volume manufacture becomes as economical as high volume production. All 'just-in-time' systems rely upon small batches and achieving target changeover times. This is often set at less than three minutes if changeover losses are to be mimimized. Standard methods of eliminating changeover loss are:

- Meticulous work preparation ensuring the materials, tooling fixtures are correct.
- Equipment preparation, test runs for machine warming, conducting routine planned maintenance–cleaning the machines, tightening bolts and correcting obvious faults.
- Establishing operational standards identifying equipment adjustments that have to be made in particular circumstances.
- Developing innovative methods for changing tooling, jig, fixtures, dies, etc.
- Agreeing critical factors that impact machine changeover times.
- Reviewing the impact of assembly line processes on the supply chain.

The team responsible for the steady state design must also consider:

- Range of products to be manufactured within a cell, identifying items that may cause changeover difficulties and propose methods for resolving problems.
- Possible standardization of product features and use of common parts.
- Common fixtures and tooling, minimizing the items to be changed.
- Tasks to be performed on the machine and tasks that can be addressed prior to the changeover event.
- Equipment needed to assist or automate the changeover process.
- Machine modifications to simplify the changeover process–manifolds, eliminating bolts, introducing standard pallets, etc.

Materials handling and Kanban techniques

Materials handling and moving components are non-value added activities. Consideration must be given to precise methods of transporting materials between cells and also within the cell. The most flexible method of moving parts between cells is using clean, reusable containers which protect parts from physical and environmental damage. These should be stackable and moved between cells using simple handling devices or trolleys. An operator placing components at the next workstation or filling the container ready for transfer should instigate movements within the cell.

Materials handling systems tend to be oversophisticated resulting in levels of automation and complexity that cannot be justified. Gravity fed devices have been used very effectively to transport high volume parts between cells and simple conveyors to pace assembly cells. However, a materials handling system forms a fundamental element of the manufacturing system and its specification must be considered an integral part of the design process. My experience has shown that simple methods for transporting materials and components are the most reliable and cost effective.

A Kanban, in its simplest form, is an *identification tag* that provides transportation and manufacturing instructions. The concept was derived from American supermarkets. Assembly people collect the number of parts they require from the component supply areas and leave tags showing the type and number of parts taken. This tag then triggers manufacturing cells and suppliers to replenish those items that have been used by assembly.

Fundamental concepts of a Kanban system:

- Assembly people should only draw from the manufacturing cells the items they need, at that time and in the exact amounts required.
- Manufacturing cells should only produce in quantities that have been drawn:
  - items manufactured should not be in excess of the number of Kanbans;
  - manufacturing should be performed in the sequence of the Kanbans; and
  - there should be no backlog of Kanbans.

The Kanban system will only work if the following conditions have been addressed:

- Changeover times have been attacked and reduced to an acceptable level.

- Number of parts identified in a container per Kanban are low, driving the concept of small lot production.
- Zero defects are accepted as a common objective and any defective parts removed prior to release to the assembly process.
- Processes used throughout manufacturing and the supply chain are process capable.

The *rules* for a Kanban system:

- Only good quality parts must be sent to the assembly area:
  - team take responsibility for ensuring good quality parts;
  - operatives have the authority to stop the process if quality is compromised;
  - supervisors take corrective actions to minimize defects and stoppages;
  - work is inspected by the operators to prevent defective parts being manufactured;
  - process capability is consistently monitored to confirm processes are robust and able to produce all parts within the specified tolerance band, focused around the mean dimension;
  - fool-proofing techniques to prevent careless errors;
  - machines equipped with error detection devices, designed to stop automatically;
  - all defects are analysed to find the possible root cause;
  - processes are documented and the conditions for producing good quality parts are recorded giving standard settings; and
  - changeover techniques and dedicated fixtures allow first-off parts to be manufactured correctly.
- Operational practices for the assembly process:
  - parts cannot be collected without an assembly Kanban;
  - number of parts collected must correspond to number of Kanbans; and
  - collected parts must be given an assembly Kanban and a machining Kanban released to trigger items to be replenished.
- Manufacturing operations must produce in the quantities used by assembly:
  - number of parts manufactured corresponds to number of Kanbans;
  - items are produced in order Kanbans are released;
  - backlog of Kanbans not allowed; this causes system failure;
  - everyone notified of significant changes in number of Kanbans being released; and
  - possible delays to delivering parts on time notified in advance.

- Number of parts associated with a Kanban should be as low as possible:
  - containers reduced in size to hold a minimum number of parts; objective is to reduce number of parts per Kanban to 1, 5 or 10 pieces;
  - only the required number of Kanbans to be released, maintaining control on level of work in progress;
  - Kanbans must be collected and circulated promptly; and
  - Kanban system 'fine tunes' the production plan, therefore sales departments must notify production of significant changes.
- Production levelling should be routine:
  - formalized methods of accommodating peaks and troughs established between sales and production;
  - manufacturing designed to handle known and agreed volumes and product variety;
  - flexibility a critical feature of cell design; and
  - specialized groups of machinery, tooling and fixtures dedicated to specific components, but with a capability of switching parts and associated tooling between cells.
- Stabilize the process:
  - confirm it is capable of producing good quality parts in allocated cycle times;
  - process can reach designed output level with zero defects using documented standard working methods;
  - reduce cycle time at appropriate intervals;
  - improve the process through continual improvement programme;
  - remove wasteful and non-value added activities; and
  - opportunities to modify it and reduce manning levels examined at regular intervals.

Nagare principles for low stock one-piece flow

Nagare principles should be considered for manufacturing cells to aid the smooth flow of materials through multi-operational production equipment. The system is designed using process flow analysis to produce non-stock, one-piece flow. The basic principles are:

- Materials flow evenly through the system with the tools and support equipment in designated places to maintain even production.
- Team members walk, progressing round the cell while the machines operate unattended, stopping on completion. The last operation is adjacent to the first, with operations arranged in the correct sequence.

- Machines are attended sequentially with the material flowing towards the finishing operations.
- Output from the cell is governed by the number of people walking around the machines; output is matched to the rate products are required.
- Manufacturing cycle time designed to correspond to customer demand pattern.
- Team members operate all the machines, including inspection routines, taking full advantage of multi-skilled working.
- Transfer quantities limited to what can be carried in two hands.
- Line designed for rapid changeover to accommodate an identified range of products.
- Plant comprises simple proven processes with narrow machine widths to reduce distances materials travel.
- Machines are reliable, easy to maintain and ergonomically designed to reduce fatigue, aiding machine changeover.
- The quality standard is zero defects with mistake-proofing incorporated into the system to prevent careless errors.
- Waste is exposed and eliminated through continual improvement of standard working practices with close attention to the flow of material within the cell.

The Nagare ideal is to achieve a consistent output performance free from defects, focusing on a regular pattern of movement by cell team members.

Standard operations

The foundation of the Toyota production system and Nagare principles is to develop standard operations combining machines and human endeavour to achieve effective production methods. Standard operations depend upon three factors:

1. Cycle time or TAKT time (*TAKT* = German for tempo) – is the time calculated on plant availability and customer volume requirements to produce a defect-free part. The cycle time is calculated for each stage of the process to synchronize operations to match customer demand.

Cycle time = number of working hours/production volume required in period

Daily production rate = one month's production volume/number of working days

2. Work sequence – the pattern of events required to produce good quality parts, not necessarily the order parts flow through the production process.
3. Standard work in progress – inventories needed at the designated workstations to allow standard operations to be performed in a specified way.

A method for establishing standardized operations (Table 3.2):

- Analyse the machining operation for each part, using time-based process flow analysis:
  - establish how the materials will flow through the proposed line;
  - undertake a process path analysis to examine the sequence of operations;
  - review the information looking for ways to eliminate waste;
  - establish modified process paths for the family of components; and
  - create new machine layouts to minimize material movements.
- Create a chart identifying the machining capacity for each part:
  - order of processing;
  - description of the process;
  - machine reference;
  - time required for each operational stage:
    - ◊ time required to complete the manual operation (excluding walking time);

Table 3.2 Example of time-based machine capacity analysis by part

	Description	*Machine ID*	*Average time (sec)*			*Cutting tools*		*Process capacity*
			Manual	*Automatic*	*Total*	*Tools to be changed*	*Time required*	
1	Insert part and machine face	L 10	6	50	56	C-100 M-166	80 sec 70 sec	
2	Drill holes	DR 20	6	42	48	D-50	15 sec	
3	Insert screws	TP 20	6	22	28			
4	Quality check		10		10			
	Total		*28*				*165 sec*	

◊ automatic cycle time when the machine runs unattended;
◊ time taken to change cutting tools and fixtures;
◊ process capacity or the number of items that can be made in one day based upon the shift pattern, number of hours worked, Kanban size and number of tool changes needed to manufacture range of parts.

Process capacity would depend upon the number of parts in the Kanban, if parts manufactured individually and cell teams worked two eight-hour shifts, then the daily process capacity based upon the longest operation would be:

$$\text{Daily process capacity} = \frac{8\text{ hours} \times 2\text{ shifts} \times 3600\text{ sec}}{56\text{ sec} + 150\text{ sec}} = 280\text{ per day}$$

(Drill change performed in cycle: 28 sec total manual time + 15 sec changeover is less than the longest cycle time 56 sec.)

If the Kanban size increased to 100 parts:

$$\text{Daily process capacity} = \frac{8\text{ hours} \times 2\text{ shifts} \times 3600\text{ sec}}{56\text{ sec} + (150/100)} \cong 1000\text{ per day}$$

- Draw up a standardized operations chart that plots the activities against the time needed to complete the tasks:
  - record the order of the operation;
  - describe the operation;
  - record the time for manual operations and automatic cycles;
  - plot the operation time using:
    ◊ solid lines – manual operations;
    ◊ dashed lines – automatic cycle time;
    ◊ curved line – walking to next station;
  - record the total cycle time;
  - use the machine capacity chart (see Chart 3.3) to determine the total manual operation time ensuring that it is less than the cycle time for the longest operation and assign required number of workers;
  - check to see if the combination of operations is feasible; making necessary adjustments to the time allotted for each task;
  - repeat the chart using an optimum walking time; and
  - check workload coincides with the total cycle time for optimal performance.

Chart 3.3 Typical time-based standard operations process sheet

	Description	Time – sec		Process time map – sec (20, 40, 60)
		Manual	Auto	
1	Take material	4		
2	Remove and mount workpiece	6	50	
3	Drill holes	6	42	
4	Tap threads	6	22	
5	Measure screw	10	–	
6	Stack item	4	–	
	Total	*36*		

(*Note:* manual time increased due to additional tasks.)

- Prepare and test the standard operations sheet (Fig. 3.1):
  - number each area of the machine layout drawing in the order that the machines will be used; join the operations with a solid line and a dashed line where the operation goes back to the beginning of the process;
  - place a ◆ on machines/processes that require a quality check;
  - place a ✚ on the machines/processes that require safety precautions;

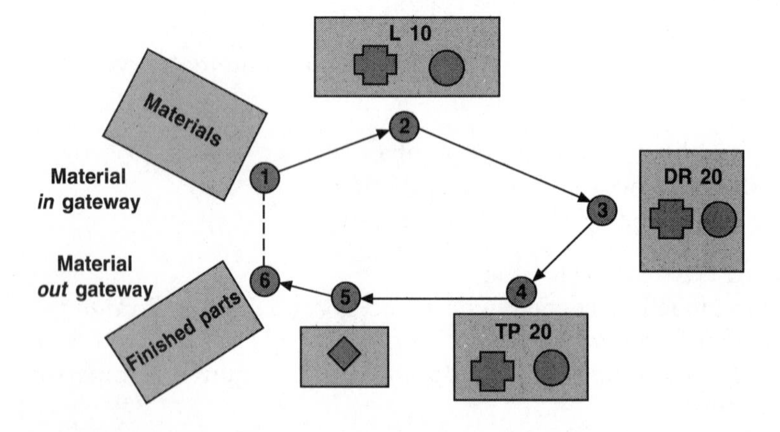

Quality check	*Safety precaution*	*Work in progress*	*Level of inventory*	*Cycle time*	*Total time*
◆	✚	●	5	56 sec	60 sec

Figure 3.1 Standard operations diagram for a U-shaped cell

- place a ● at the processes that require a standard number of work in progress items with the number of pieces required;
- indicate the cycle time; and
- note total hours required to produce the required quantity.

Bottlenecks

Bottlenecks are important in all manufacturing systems, because they are a major constraint on performance. They can usually be identified by high levels of work in progress where parts queue to pass through the workstation. The bottleneck limits the overall performance of the cell impacting the capacity of the cell, work in progress, lead times, responsiveness and the overall cost of manufacture; it may be caused by equipment, a group of people or even an item of tooling. It is important to identify them at the design stage as they ultimately limit the overall capacity of the factory. In some situations the bottleneck may move depending upon the volume and mix of products being manufactured. This situation requires careful manufacturing system design and allocation of adequate resources to maximize the potential output. Possible bottleneck operations should be given serious consideration when allocating capital investment and must be the focus of continuous improvement activities, as reductions in cycle times or changeovers provide immediate benefits, increasing available capacity and reducing work in progress. The management of bottlenecks requires rigorous schedules to deliver full capacity; they should be protected from all possible unplanned events, using increased levels of work in progress if necessary, as stoppages result in lost output that may never be recovered.

I have not intended to give an in-depth exposition of Japanese manufacturing methods, but an appreciation of the concepts and levels of precise detail used to design their manufacturing systems. Many books have been published on the subject including *The Toyota Production System. An Integrated Approach to Just in Time*, Yasuhiro, M. (1998), London: Chapman and Hall; these give more detailed information on the methods used. This rigorous attention to detail when designing the manufacturing system goes a considerable way to explaining why the Japanese have dominated manufacturing industry in recent times. The traditional approach of allowing the manufacturing process to evolve without senior technical management involvement in the way systems are designed has resulted in the Japanese becoming very serious competitors in many market sectors worldwide.

Final comments on the steady state design

At the end of the steady state module and cell design process, manufacturing processes and all significant cell resources should have been defined, allowing the cell performance for lead time, stock levels and operating costs to be calculated. The structure will normally comprise of a combination of different modules, for example:

- Runner modules – core original equipment products.
- Customer-focused modules – original equipment supplied to major accounts.
- Repeater modules – high variety, low volume products.
- Stranger modules – support aftermarket or product development items.
- Complex machining or process modules – parts manufactured on capital intensive equipment that cannot be justified in more than one module.
- Heat treatment and processing modules.
- Repair and overhaul modules.

The specific number of modules will be dependent upon the type of business, customer requirements and the manufacturing strategy being adopted, but the structure must now be *formally agreed* before proceeding further with the manufacturing design process. An initial plant layout should be prepared showing the amount of space required for the new manufacturing system, location of plant and equipment, flow of materials and any constraints with the proposed location that may exist within the factory.

> *A team of people who understand how the manufacturing process should operate, and designing a process to meet those requirements, ultimately must create a more cost-effective solution, compared with traditional manufacturing systems which have evolved, taking little or no account of actual business needs.*

The steady state design of manufacturing cells is a complex process, requiring a wide range of factors to be considered. No standard solution exists and each situation must be analysed separately. Initial resource requirements will have been calculated in considerable detail, but the design is still based upon stable, average values that may not represent actual factory dynamics. Therefore, further work is required to address support functions and test the robustness of the design against variations, including unplanned events that may occur within the business environment.

Chart 3.4 is an input/output diagram which lists items that should have been confirmed on completion of the module/cell steady state design phase.

Chart 3.4 Input/output analysis for the module/cell steady state design

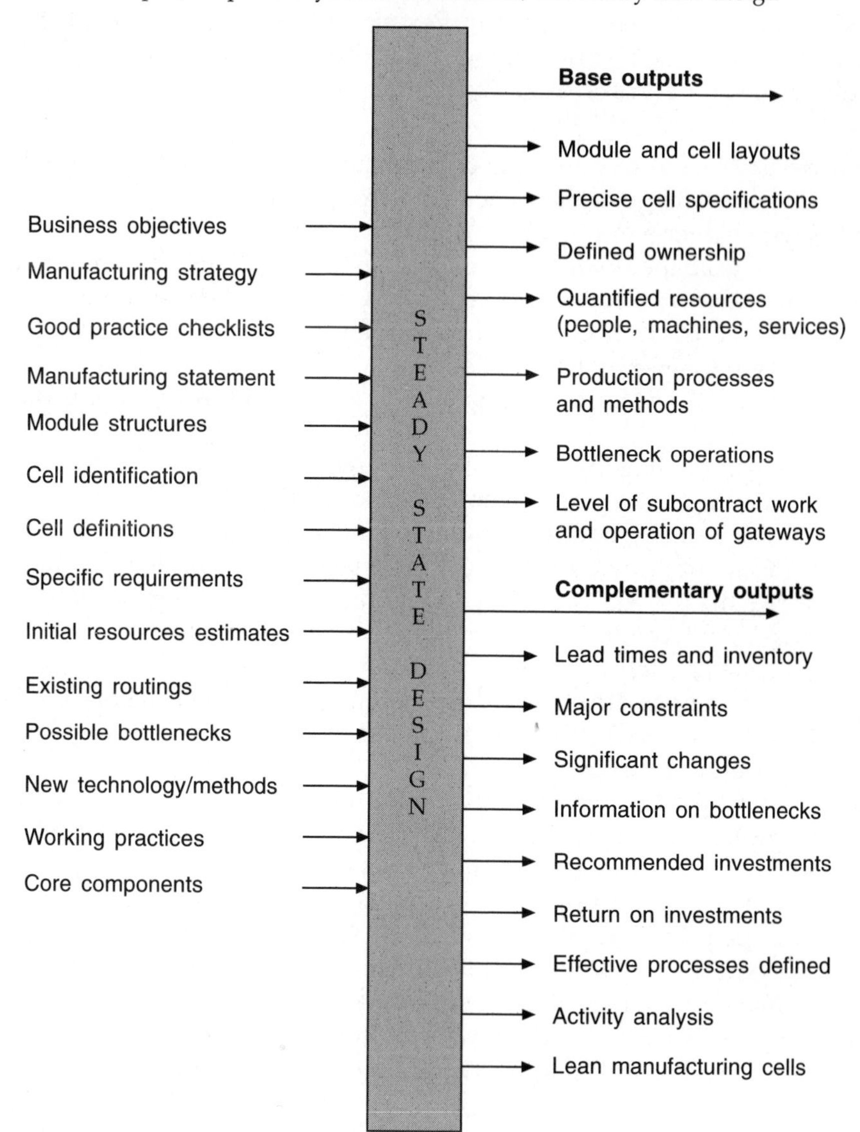

Job design

Once the steady state manufacturing module and cell design has been completed, the next stage is to define the job roles needed to make module and cell teams work effectively. The number of manufacturing people needed to perform assembly and test activities or run the machines will have been identified from the steady state analysis, but these need translating into actual job requirements. The nature of cell teams places great emphasis on people being flexible and willing to expand their range of skills. Cell team members must be prepared to undertake the following type of tasks, after appropriate training:

- Assemble and test a family of products manufactured by the cell.
- Operate a range of manufacturing equipment.
- Change over equipment for different products.
- Assist team members to complete specific operations when required.
- Move around a number of machines and keep them running.
- Verify products meet the quality standards expected by the customer.
- Move parts between workstations.
- Clean components, protecting them from physical and environmental damage.
- Perform manual in-process or finishing operations.
- Collect materials/components needed for the job.
- Kit the tools needed to manufacture or test products.
- Conduct first-line preventive maintenance on equipment/tooling.
- Verify the process capability of equipment/measuring systems.
- Plan own work to meet commitments made to customers.
- Operate specialist processes needed within cell.
- Modify and create standard operational process instructions.
- Introduce agreed process modifications to improve team performance.
- Train other team members.
- Support the team to meet its commitments.
- Complete documentation on quality performance, production levels and other key measures of performance.
- Contribute to continual improvement activities.
- Change processing fluids and check they are correctly maintained.
- Support the environmental health and safety policy.
- Remove waste materials from work area.
- Keep working area a clean safe environment to work in.
- Work flexible shift patterns and overtime when necessary.

This list of job requirements highlights the major changes that have to be made in the way people work. In my experience, the overall rate of change is governed by the amount of training needed to equip people to take on these broader roles and responsibilities.

Activities to be undertaken within the cell will have been established during the steady state design phase, and range of skills determined by the type of cell and range of tasks performed, but job requirements must embrace the following:

- Assembly and test teams will:
  - handle a range of products;
  - operate a range of assembly processes and test equipment;
  - accommodate changes in product mix as routine;
  - perform all activities, including final conformance assurance, rectification and verification; and
  - pack products in containers ready for use by the customer.
- Component manufacturing teams will:
  - produce a range of components in a way that does not compromise the cost-effective manufacture of high volume parts;
  - undertake all the processes needed to complete components and subassemblies, taking full responsibility for supplying quality assured products directly into the assembly module;
  - be responsible for confirming the capability of processes; and
  - perform first-line maintenance on equipment.

The next task is to determine the additional functions needed within the modules/cells to support production processes.

Support services and indirect resources

The objective of modular manufacturing is to establish local ownership, but people can only be held directly responsible for activities under their control. It is desirable therefore to incorporate necessary support activities within the module, or cell, ensuring team members have the resources needed to take responsibility for delivering quality products on time. Module support services to consider include:

- Management and team supervision:
  - management supervision structure and reporting relationships;
  - management needed on different shifts; and
  - working team leaders, supervising self-directed work groups.
- Technical support:
  - quality assurance and conformance;
  - manufacturing engineering to resolve day-to-day production issues;
  - technical support to resolve product-related items;
  - production scheduling, planning day-to-day activities based upon availability of resources;
  - maintenance support and facilities management;

- material procurement ensuring materials are available on time;
- work in progress tracking and performance measurement; and
- continuous improvement teams.

- General support:
  - stores and logistics;
  - servicing machines, changing fluids and disposal of waste substances; and
  - cleaning gangways and general areas.

These support functions are non-value added but must be incorporated into the team's activities. However, it may be necessary to initially assign people to undertake specific tasks, because until teams have been trained in the broader aspects of their new roles (operating to standardized procedures using process capable equipment), it is unwise to make too many cost savings on overhead support. Removing existing support activities prematurely can result in a serious loss of skills, which could threaten the viability of the business. Once the teams are trained and proficient in their broader roles it is much easier to operate with minimal overheads.

> *It is worth remembering that the Japanese manufacturing revolution took many years to accomplish!*

The management resources and site support needed outside the modules are dependent upon other business activities on site, but consideration should be given the following areas:

- Management of the site:
  - operational management structure and reporting relationships;
  - site supervision required on different shifts; and
  - management of specialist activities and services.

- Technical and staff support:
  - customer satisfaction, quality assurance and conformance;
  - manufacturing engineering, supporting design for manufacture;
  - manufacturing systems redesign and factory reorganization;
  - industrial engineering, production equipment and tooling;
  - project management of change programmes;
  - technical support to resolve product design issues;
  - master production scheduling at each of the three levels;
  - planned maintenance support, plant refurbishment and installation;

  - facilities management for the site;
  - purchasing and contract negotiation;
  - procurement and supplies management;
  - information technology and systems;
  - coordination of continuous improvement teams;
  - personnel and industrial relations;
  - skills assessment and training; and
  - environmental health and safety.

- General support:

  - stores management;
  - logistics, transportation and distribution;
  - spares repairs and aftermarket support; and
  - site services.

The organizational structure should be simple with clear ownership and a maximum of four levels from the general manager to shop floor. The overhead support must be held to strictly controlled management guidelines with people taking responsibility for several tasks.

Quality system

Quality procedures

A quality system must be designed to satisfy business requirements, conforming to the ISO 9000 series of standards. A system of quality management within the company must allow customers to evaluate suppliers and compare their ability to reach acceptable levels of quality and reliability. The principles of the ISO standard are based upon adopting formal processes and standardized procedures to achieve consistent, good quality, not the detailed activities underpinning a process.

Basic requirements for quality system design is to identify supply-chain elements which have an impact upon product quality or customer service, and to prepare documented procedures that facilitate operational consistency in a planned manner. Therefore the ISO 9001 standard is based upon establishing a formal quality procedure covering the following supply-chain activities:

- Purchasing control, ensuring products conform to specified requirements.
- Identification of products and traceability, linking them to specifications and work orders.

- Control of manufacturing processes with work instructions and documented procedures.
- Control of specialist processes, including environmental requirements.
- Methods for inspection and testing product conformance:
  - receiving inspection;
  - supplier quality assurance;
  - in-process monitoring and control;
  - product monitoring and conformance verification;
  - final inspection and performance testing; and
  - inspection records and documentation.
- Inspection of measuring and test equipment, calibration, capability and condition.
- Methods of inspection and recording the test status of products in the factory.
- Control of non-conforming products and methods of identification.
- Corrective actions to rectify defects.
- Handling, storage and delivery of products to customers.
- Quality records and methods of retrieving/controlling information.
- Internal quality audit procedures.
- Training requirements and skills needed for operating to the quality procedures.

Documentation control is a critical aspect of the ISO quality process:

- All quality and operational instruction documents must be approved and authorized prior to release.
- Current issue must be used for manufacture.
- Documents must be systematically updated.
- Obsolete documents must be removed to prevent possible error.
- Changes to the product, process or procedures must be formally recorded.
- The current issue of documents must be easily identified and available.

The actual preparation of formal quality procedures is a considerable task, particularly when taking account of the training required to ensure the workforce's skills are compliant with written procedures. They cannot be prepared until the supply-chain process has been defined, but it is important to scope the requirements of the quality system, ensuring that supply-chain processes are aligned to the company's quality policy. Cell teams should be involved in writing these procedures as they are expected to implement and use them. Audits must also be introduced to verify that procedures are being followed implicitly. If written in a bureaucratic style, implementation could be extremely difficult! To gain ISO accreditation the system must be finally audited by an approved third party agent, confirming that procedures are being followed correctly.

Cost of quality

The cost of quality is an important factor in the supply-chain process. Targets should be established for a series of quality-related performance measures, which can be used to quantify cost savings and benefits from the new supply-chain and manufacturing processes (see Table 3.3).

Personnel policies

The supply-chain and manufacturing system design will not be successful without people understanding the need to change and be fully involved in the process. Making changes to the way people behave and work is a challenging but also a most rewarding management task. An employee development route map (Chart 3.4) needs drawing up, guiding the workforce through the transition. An environment of mutual respect

Table 3.3 Typical quality measures to determine the cost of quality

Quality measures	*Definition of measure*	*Performance*	
		Current	*Target*
Cost of quality			
Failure costs	The burden of not performing work right first time	$	$
Internal	Costs associated with scrap and rectification including both labour and materials; and the proportion of product introduction costs needed to modify designs after being released for production	$	$
External	Warranty claims and costs of rectification including product substitution programmes		
Quality assurance costs	Financial cost of operating the quality assurance system including calibration, process capability, preparation of documentation and training	$	$
Quality impact on profit	Failure costs expressed as percentage of trading profit	%	%
	Failure costs expressed as percentage of sales	%	%

Chart 3.4 Employee development route map

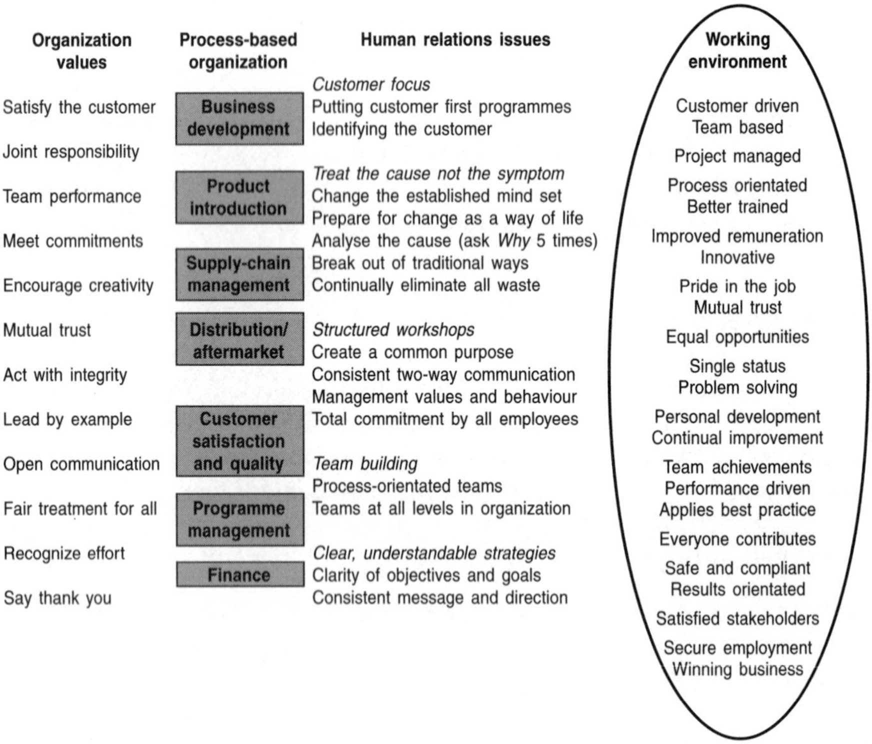

Organization values	Process-based organization	Human relations issues
		Customer focus
Satisfy the customer	Business development	Putting customer first programmes Identifying the customer
Joint responsibility		
		Treat the cause not the symptom
Team performance	Product introduction	Change the established mind set Prepare for change as a way of life
Meet commitments		Analyse the cause (ask *Why* 5 times)
	Supply-chain management	Break out of traditional ways
Encourage creativity		Continually eliminate all waste
Mutual trust	Distribution/ aftermarket	*Structured workshops* Create a common purpose
Act with integrity		Consistent two-way communication Management values and behaviour
Lead by example	Customer satisfaction and quality	Total commitment by all employees
Open communication		*Team building* Process-orientated teams
Fair treatment for all	Programme management	Teams at all levels in organization
Recognize effort		*Clear, understandable strategies*
	Finance	Clarity of objectives and goals
Say thank you		Consistent message and direction

Working environment

Customer driven
Team based
Project managed
Process orientated
Better trained
Improved remuneration
Innovative
Pride in the job
Mutual trust
Equal opportunities
Single status
Problem solving
Personal development
Continual improvement
Team achievements
Performance driven
Applies best practice
Everyone contributes
Safe and compliant
Results orientated
Satisfied stakeholders
Secure employment
Winning business

Reference: *Plan to Win* (Garside 1998)

must be established that promotes successful teamwork, where people are proud to be members of a successful team. The proposed supply-chain and manufacturing systems design will be based upon teams of people performing a number of inter-related tasks with everyone accepting the challenge of training in new skills and continual personal development.

> *The magnitude of this change must not be underestimated; the management team must create a level of trust allowing the workforce to feel confident in making the transition. It must also appreciate the radical changes needed in its own behaviour, management style and organizational structure.*

Employee framework

A framework identifying the characteristics of the working environment must support the introduction of new working practices:

- No false barriers to achieving full job flexibility, except one's own ability.
- Everyone having the same basic terms and conditions of employment, as allowed by national legislation.
- Appropriate training being available to advance knowledge and skills.
- Remuneration systems rewarding the acquisition and application of skills.
- People being encouraged to achieve their full potential.
- Recognizing people giving leadership and accepting responsibility.
- Relationships founded upon mutual respect and trust.
- People feeling involved, committed, and seeking opportunities for improvement.
- Teams accepting ownership and responsibility for delivering tasks on time.

Skills audit

The size of task involved in training people to acquire broader skills and to change the organization culture can be assessed using a self-assessment skills audit and attitude survey. A range of views (which must be obtained confidentially) include:

- Flexibility.
- Present ownership, control and accountability.
- Training and personal development.
- Terms and conditions of employment.
- Collective reward systems.

People should be asked to provide a yes/no answer to a number of questions, writing comments if they are unable to state a clear position. Individual responses to the survey must be kept confidential and analysed to provide collective representative views for the management team to consider. The questions should be structured to elicit a realistic insight into how the workforce feel about the prospect of change and any important issues that must be addressed during the system design process.

Flexibility

- Are there any false barriers to you achieving full job flexibility?
- Would you be prepared to perform any machining, assembly test or other production process within a team environment?
- As a team member, would you be prepared to carry out production

engineering activities to overcome unplanned plant breakdown, improve process capability and eliminate waste?
- Are you computer literate?
- Are you able to programme numerical machines and test equipment, set up fixtures and install tooling?
- Are you prepared to perform first-line maintenance and general cleaning activities unassisted, but calling for assistance when needed?
- Will you take responsibility for the quality of your own work, completing quality control documentation, monitoring process capability, using measuring systems and maintaining records specified in the quality procedures?
- Will you operate the manual or computerized work scheduling systems, ensuring parts and products are available on time to meet customer schedules?
- Will you fulfil any administrative system requirements associated with the module and/or cell?
- Are you prepared to organize materials for members of your team?
- Are you able to create production layouts and tooling requirements for jobs?
- Will you move between teams or into support areas when required, to overcome particular business circumstances?
- Have you experience of:
  - supervision?
  - operating all cell processes?
  - product applications or development engineering?
  - scheduling?
  - inspection?
  - production engineering?
  - maintenance?
- Are you prepared, subject to normal protocol, to move freely between shifts when business circumstances require?

Present ownership, control and accountability

- Do you work in a team environment?
- Do individuals and team members take responsibility for their collective tasks, including self-supervision, quality assurance, scheduling and general housekeeping?
- Do you take personal responsibility for setting up machines and verifying first article inspection?
- Are you responsible for your own quality of work, confirming capability of processes and measuring systems and, where appropriate, final inspection?

- Do you organize your own activities and standards of work?
- Are you responsible for recording your own attendance hours and job bookings?
- Does your team organize its own materials, tools and equipment?
- Does your team know customer requirements and schedule its workload to respond to customer or business commitments?
- Is your team responsible for organizing who performs particular tasks?
- Does your team communicate with other shifts and supervisors?
- Does your team communicate directly with support activities/departments?
- Has your module team well-established continuous improvement groups?

Training and personal development

- Do training facilities exist to provide you with the skills you need to work in a team environment?
- Are you encouraged to undertake training with the support mechanisms in place to help you achieve your desired potential?
- Do you know if the business has conducted a thorough skills audit and matched it against business requirements?
- Do you know if the business has identified your skills gap and discussed your requirements to support an evolving organization?
- Do you have a personal training and development plan?
- Has the business provided you with the necessary skills training in business awareness and control systems?
- Are facilities available for you to train in the necessary computer skills?
- Has the company provided dedicated resources to underpin the requirements of team working, problem solving, recording information and new skills training?
- Has the company introduced assessments, to ensure you are proficient once your training has been completed?
- Has the company involved a third party assessment for training to provide accreditation?
- Have you been involved in providing training for your colleagues?
- Are you able and prepared to deliver training to your colleagues?

Terms and conditions of employment

- Has the company introduced a single status environment where everyone has equality in basic terms and conditions of employment?

- Are overtime premiums common across the business?
- Are the premiums paid on the full rate?
- Are sickness benefits common?
- Are holiday entitlements common?
- Do common policies exist for controlling payments for authorized absence?
- Has the company single status restaurant, parking, toilets, etc.?
- Do you have time and attendance clocks?
- Has the company equal terms for lay-off provisions?
- Is remuneration linked to personal skill acquisition and application?
- Has the payment structure been integrated into a single structure?
- Are payments by monthly credit transfer?
- Does the business operate an individual performance review and related pay system?

Collective bargaining arrangements

- Would introducing team working practices necessitate a collective bargaining process?
- Can you describe an employment vision in terms that satisfy both the employees and trade unions?
- Do new requirements need incorporating into an agreement that requires the acceptance of local trade unions?
- Do they have to be negotiated with each trade union separately?
- Do you have a single bargaining unit?
- Do you understand long-term implications for employment levels?

Leadership values

Acceptance of fundamental change can be facilitated by the management team providing a statement on its leadership values. People will usually respond positively to the need for change and accept the impact it will have upon their lives if they understand the reasons. It is important that senior managers provide leadership, making time to communicate continually with everyone in the business. Such values should be established collectively by the senior management team and the workforce. Only those values that can be upheld should be included in a statement of intent, which may include:

- Take pride in team achievements.
- A belief that teams can achieve more than individuals.
- Decisions must be based upon integrity.
- Mutual trust and respect for the individual.

- Open two-way communication that invites constructive discussion.
- Sharing information and business performance measures.
- Leadership by example.
- Encouraging people to take calculated risks.
- Decisions will be made for the benefit of the majority.
- Recognizing an individual's effort.
- Taking the opportunity to say thank you.

These statements must be given full and careful consideration. They have huge potential to unite the workforce and ease the change process, but it must be noted that if people feel betrayed, perceiving that managers are failing to live up to these values, the consequences can be extremely damaging.

Support function design

Supplies module

The structure of the purchasing organization and its role within the company are dependent upon the business requirements. It must be designed to support the supply-chain process, providing the following four key purchasing activities: sourcing, materials planning, inventory management and procurement. These should be performed by a materials group, operating as an integral part of the supply-chain process. If a business has a significant product introduction programme, sourcing and supplier development activities could be integrated into the product introduction management team.

The external environment and its impact upon a business have direct implications on the purchasing role of the company:

- Markets are becoming global.
- Opportunities to source from international suppliers are increasing.
- Partnerships and preferred suppliers are an established way of conducting business.
- Reciprocal trade agreements are of growing importance for winning business.
- Customers and competitors are adopting collaborative agreements, sharing risks and revenues.
- Decreasing acquisition costs, stringent quality requirements and improved delivery performance are demanded by customers.
- The rate of technological change is ever increasing.

These factors have a significant influence upon product technology, manufacturing strategy and purchasing policies for companies, resulting in:

- A continuing trend to increase bought-out content of products.
- Greater importance on component total acquisition costs as a key factor in business profitability.
- Increased rate of change for product technology, leading to shorter product life cycles needing greater technical understanding linked to purchasing expertise.
- Logistics and material management techniques becoming core business competencies providing competitive advantage from the successful integration of supply-chain interfaces.
- Increased outsourcing of non-core components, changing manning levels and releasing factory space.
- Reassessment of skills needed to manage the business, with skills shortages rectified through training and recruitment.

All these factors impact the business, but the most important aspect for the supply-chain steady state design is to determine the operational requirements that is the size of team needed to take responsibility for materials planning, inventory management and procurement. These processes are fundamental to ensuring regular supplies of good quality products, on time, to support internal manufacturing processes. This can be achieved through creating a *supplies module* supporting supply-chain operations, servicing both manufacturing cells and assembly modules.

The main tasks for the supplies module include:

- Agreeing the master production schedule, confirming the capacity from the supply base to achieve the projected production volumes in the medium and long term.
- Confirming the production schedule with suppliers ensuring materials will be available within the lead time to meet the production plan.
- Establishing level of demand and frequency of deliveries to satisfy the production plan and expenditure constraints agreed with the finance manager.
- Ensuring that all materials and components are to specification and available on time to meet the production plan.
- Working with suppliers to reduce their lead time needed to fulfil orders.
- Verifying that the lead time on the material requirements planning system matches the lead times agreed with suppliers.

- Setting up an effective interface with suppliers to address queries, changes to requirements or quality issues.
- Implementing supplier rating systems that monitor individual performance against key parameters including a cost index, quality plan, delivery to schedule and perceived customer satisfaction.
- Establishing a list of preferred suppliers who consistently achieve the required performance level and can be allowed to deliver products directly to production without formal inspection or verification.
- Providing assistance and support to suppliers and introducing improvement programmes to meet the performance levels expected from preferred suppliers.
- Working with suppliers to reduce the total acquisition cost, securing savings that can be obtained from working in partnership, removing non-value added tasks in the process.
- Authorizing payments for goods received as per the terms and conditions agreed in the contract.

These activities focus upon servicing manufacturing processes and are crucial to achieving an effective supply chain but they cannot be divorced from other professional purchasing activities: strategic sourcing, price negotiation and the supplier development programmes. These are crucial tasks needed to meet financial business commitments and support the product introduction process. Broader purchasing roles should be considered in relation to the overall business requirements and also the impact a professional purchasing team could make upon business profitability. Considerable benefits can generally be gained from combining individual business requirements to establish company-wide contracts for similar production items, subcontract services and also non-production commodities such as energy, travel, computer hardware, office supplies and company vehicles.

Administration

Additional administration support required for manufacturing modules should be minimal, with cell teams directing their own activities and accepting responsibility for quality. However, self-directed work groups still require some administrative support for managing external activities and interfacing with other business areas. Consideration must be given to the level of support required and the tasks needed to coordinate activities. These may include:

- Negotiating the production schedule and production plan for the team.

- Confirming the plan is achievable using rough-cut capacity planning tools.
- Identifying resource requirements to meet production volumes.
- Recommending actions needed to match resources to demand.
- Identifying skill gaps in the team and recommending training requirements.
- Organizing and providing 'on-the-job training' for the team.
- Maintaining records on training/skills of individual team members.
- Organizing short stand-up meetings to provide regular feedback on achievements/concerns.
- Coordinating improvement groups, seeking resources for implementation.
- Taking responsibility for determining root causes of problems.
- Scheduling planned maintenance into the production plan.
- Liaising with other areas and obtaining support to resolve problems outside the team's remit.
- Controlling the flow of work into and out of the cell and maintaining information on its movements.
- Coordinating the availability of materials, machines and people to deliver the production plan.
- Controlling consumable items used by the module.
- Establishing and maintaining critical measures of performance for the business, which can be influenced by the module and cell teams.
- Conducting cell audits to confirm that standards are maintained.
- Verifying processes and measuring systems are all process capable.
- Participating in preparing cell procedures ensuring they are appropriate, documented and rigorously followed.
- Managing human relations/personal issues.
- Monitoring cell performance against targets set in the business objectives.
- Maintaining records, documentation and work instructions, ensuring they are current and correct.
- Checking the calibration of measuring systems and instruments used in the production process.
- Confirming quality procedures are followed and appropriate documentation completed for customer records.
- Instigating corrective actions to resolve problems that could impact customer satisfaction, quality delivery, or production costs.

The ownership of these tasks will depend on the management structure required to administer the supply-chain process across the business, but at this stage consideration should be given to those elements that need to be incorporated within the module or cell structures.

Logistics

The method of transporting materials between different manufacturing areas has a significant impact on damage caused to components due to movement between workstations and the level of capital investment required. Experience has shown that for low or medium volume manufacturing, introducing simple containers that separate components preventing physical contact provides a cost-effective method for handling components. These containers must be clean and stackable, allowing no contact between components. A method of transporting containers between cells must also be identified, for example by hand-carrying containers or by using an appropriate trolley for larger parts. This underpins small batch manufacturing, promoting personal contact between cells and sometimes acting as the Kanban, triggering production.

High volume facilities requiring quantities of similar components may benefit from the installation of automated or gravity-fed conveyors; such systems are used to link machines and/or assembly processes, providing a continual flow of component parts that can be loaded onto a machine or installed into a final assembly. These operations can be performed manually or by automated 'pick and place' units depending upon the level of automation that can be financially justified. Handling systems, unless precisely specified and carefully designed to meet specific requirements, can be too complex, introducing sophisticated equipment that may be unreliable and cost more than machines performing value added operations. They must always be fitted with error detection devices able to stop the process or automatically remove faulty components from the line, alerting people to any problems.

Investment in more sophisticated handling equipment is usually justified for transporting large or heavy items in high volumes. Such situations usually require a bespoke solution that may use track conveyors or automatically guided vehicles to help maintain a steady flow of work.

At this stage in the design process the critical requirement is to identify the level of mechanization and capital expenditure required to move materials around the factory and where necessary out to subcontractors. Transporting materials is a non-value activity and if it can be achieved by combining operations or establishing single piece flow through a group of machines placed closely together, then *elimination* is a more elegant solution than investment in an automated process.

Facilities supporting production

Particular industries require specialist facilities to support the manufacture of high quality core components. The equipment needed to provide these in-house facilities must be technically evaluated and financial investments fully justified based upon the strategy requirements, effective utilization, production volumes and capital expenditure levels identified in the company business plan. They are often expensive to purchase, install, operate and maintain but are often fundamental to establishing a reliable and robust process. Types of facilities to be considered could include:

- Purchase or refurbishment of heat treatment furnaces.
- Plating and surface modification processes.
- Water treatment and disposal.
- Effluent treatment.
- Transportation and disposal of waste created by production process.
- Painting, coating and drying facilities.
- Clean room facilities for product assembly or component manufacture.
- Fume and dust extraction.
- Acoustic barriers and noise attenuation devices.
- Electricity substations.
- Refurbishment of facilities, floors and buildings.
- Reinforced foundations for equipment.
- Cooling water systems.
- Cranes and lifting equipment.
- Security systems.
- Loading and unloading bays.
- Storage areas and warehouse facilities.
- Storage and supply of specialist gases and liquids.
- Bulk storage hoppers for materials.
- Compressors for distributed high pressure airlines.
- Items needed for compliance with international, national and local environmental, health and safety standards.
- Wiring looms for module communication systems and computers.
- Non-solvent cleaning processes and tanks.
- Environmental cleaning and clear-up of contaminated areas.

Summary

The steady state design of the manufacturing system is a challenging and time-consuming task, requiring commitment from the management team supported by a full-time project manager leading a dedicated multi-skilled manufacturing team. The process depends upon assimilating considerable information that may not be readily available within the company, but it must always be remembered that an informed estimate is a far better basis for taking decisions than acting in ignorance. Project owners must make important decisions, using judgement and current available information, if a steady state system design is to be available

Chart 3.5 Input/output analysis for steady state design of the manufacturing system

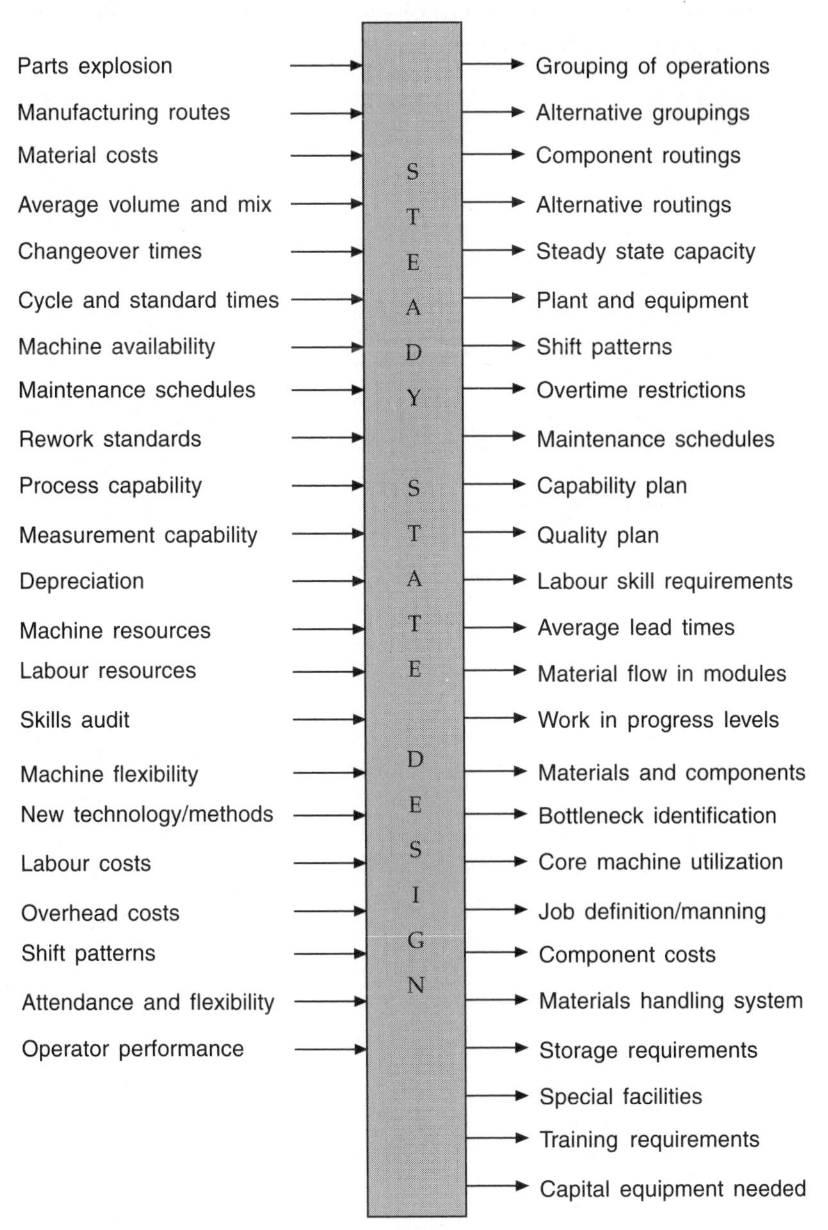

within an acceptable timeframe. The Japanese have perfected the process for designing supply-chain and manufacturing systems, using an incredible level of detail when specifying appropriate production methods. I believe this has given their companies considerable competitive advantage, particularly when linked to continual improvement programmes following implementation. The input/output analysis in Chart 3.5 summarizes the information that should have been collected and used to establish a steady state manufacturing process design.

Chapter 4

Dynamic design of modules and control systems

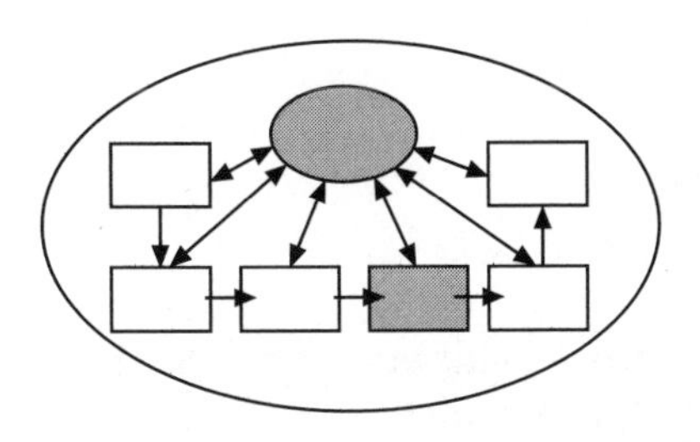

Introduction

The dynamic design stage follows steady state design, attempting to predict more accurately how the system will behave under *typical* operating conditions. Steady state design assumed that conditions remained stable and parameters were assigned average values. The dynamic analysis tests the validity of these assumptions and determines the likely impact of unplanned events. All systems exhibit transient behaviour when subject to change, but conditions that cause the greatest disruption must be understood and the system designed to react to these changes in a controlled manner. The robustness of this initial design must be tested against possible disturbances with modifications introduced to prevent failure of critical aspects because of pressure from increased production or predicable events. The objective of this phase is to allow the actual production capacity to be predicted, compared to the forecast requirements and cell specification. Any shortfall must be identified and recommendations made on how to modify the steady state design and introduce control procedures to achieve the desired output levels under a variety of operating conditions.

Factors influencing dynamic design

Manufacturing production is dynamic in nature being subject to many external and internal influences that impact performance; these may include:

- Variation to the sales volume required either immediately or medium term.
- Changes to the mix of products due to changes in demand/seasonal variation.
- Urgent jobs required to aid recovery from unforeseen events.
- Suppliers failing to meet delivery promises.
- Excessive supplier lead times.
- Shortages of raw materials.
- Quality problems with purchased items.
- Machine or plant breakdowns.
- Power failures.
- Lack of spare parts for repairing essential plant and equipment.
- Shortage of consumable items needed to complete specific tasks.
- Quality problems resulting in scrap, corrective actions and rework.
- Remanufacture of faulty items.
- Customer not providing technical information.
- Customer quality audits and corrective actions.
- Variation in operator performance.
- Absenteeism through holidays or sickness.
- Lack of skills to operate the full range of tasks.
- Training required in new production methods, quality procedures, or basic skills.
- High profile factory visits.

During the steady state design several of these areas should have been addressed, making allowances for such events. However, their impact would have been assessed in *overall* terms, not specific to unplanned events that could have a dramatically greater effect on the factory's ability to manufacture. For instance, an average machine breakdown allowance of, say, 4 per cent per year may be built into the capacity planning calculation, but if the actual failure occurs on a bottleneck machine and this is unavailable for a week, then production output and customer service will be decimated. Similarly, if an operator with specific key skills is absent, unless action has been taken to train other team members in these, the customer schedule could again be in jeopardy. Installed capacity that may be adequate under normal circumstances and average mix conditions, in practice may be inadequate when considering transient and dynamic situations. Processes can be made less sensitive to fluctuating demand/disruption by ensuring all processes are capable and increasing the flow of work through the supply chain. This can be achieved by reducing batch sizes and removing work in progress from the system. Traditional production methods employing high levels of work in progress used inventory to mask problems, making the system more tolerant to failure at the expense of being unresponsive to customer requirements, long lead times and high levels of working capital. In current competitive global markets, systems requiring excessive working capital are not financially viable.

Identification of long-term dynamic demand

Long-term variability comes from those events imposed on the business from outside, in addition to those present within the organization. External influences emanate from customers, suppliers and the general nature of business cycles that drive demand. These variations are complex, influencing both long- and short-term requirements. Product life cycles follow a classic pattern of growth, stable demand and decline. If the impact of these cycles is examined at component level for parts used in more than one product, being subject to normal fluctuations across an industrial sector, these effects combine to produce continual shifts in both total volume and mix of products required. The situation can be further exacerbated by planned obsolescence of components by suppliers, forcing more changes on product configurations and manufacturing processes. Internal influences are typically governed by management needing to increase capacity, consolidate facilities, reduce costs or marketing policies governing a particular range of products. However, full consideration must be given to a plant's capacity. Once the production requirement exceeds the volume available on bottleneck processes, manufacturing redesign and investment decisions have to be made concerning additional equipment needed to satisfy future demand, while understanding the impact of component life cycles. Plant and equipment also have finite lives; when this is no longer process capable, available capacity is also compromised.

Dynamic design tools

A study of dynamic behaviour requires complex modelling tools; for a more rigorous analysis comprehensive computer simulation models may be necessary. Computer simulation modelling packages are available from a number of manufacturing software specialists. Such models can be difficult to build, demanding considerable manufacturing and financial expertise for interpreting optimum performance. However, several important operational parameters for a manufacturing system can be mapped very effectively using spreadsheet models to perform sensitivity analysis, even though they cannot replicate random events occurring within a factory environment. It must also be appreciated that modelling techniques only assess system performance under a given set of parameters. It is the task of manufacturing engineers and the design team to ascertain these critical parameters and assign the range of values needed to give a representative evaluation.

The dynamic design process

This systematically considers features known to vary over time, having a direct impact upon criteria used for steady state design. Some modification to parameters established in the steady state design phase will be required to accommodate the transient conditions, and then be linked to a module control system capable of handling dynamic events. The items to be considered are shown in Chart 4.1.

Production volumes

The production volumes required may change from one period to the next as shown in Fig. 4.1. This fluctuating load must be analysed to establish how the manufacturing and control system will accommodate these variations.

In this example, the average production volumes required for the range of components is 6000 units per month and this figure would have been used for steady state design calculations. It now requires a more detailed analysis, taking account of changes that occur between periods. In this situation several alternative solutions need evaluating depending upon existing supply-chain process constraints.

Possible scenarios for handling the variations in demand:

- Level schedule the production to 6000 units per month, maintaining a predictable load and carrying additional stock to meet peak demands.
- Adjust the manning and overtime levels to meet output requirement.

If the situation is constrained by limiting bottleneck operations to the

Chart 4.1 Dynamic design process

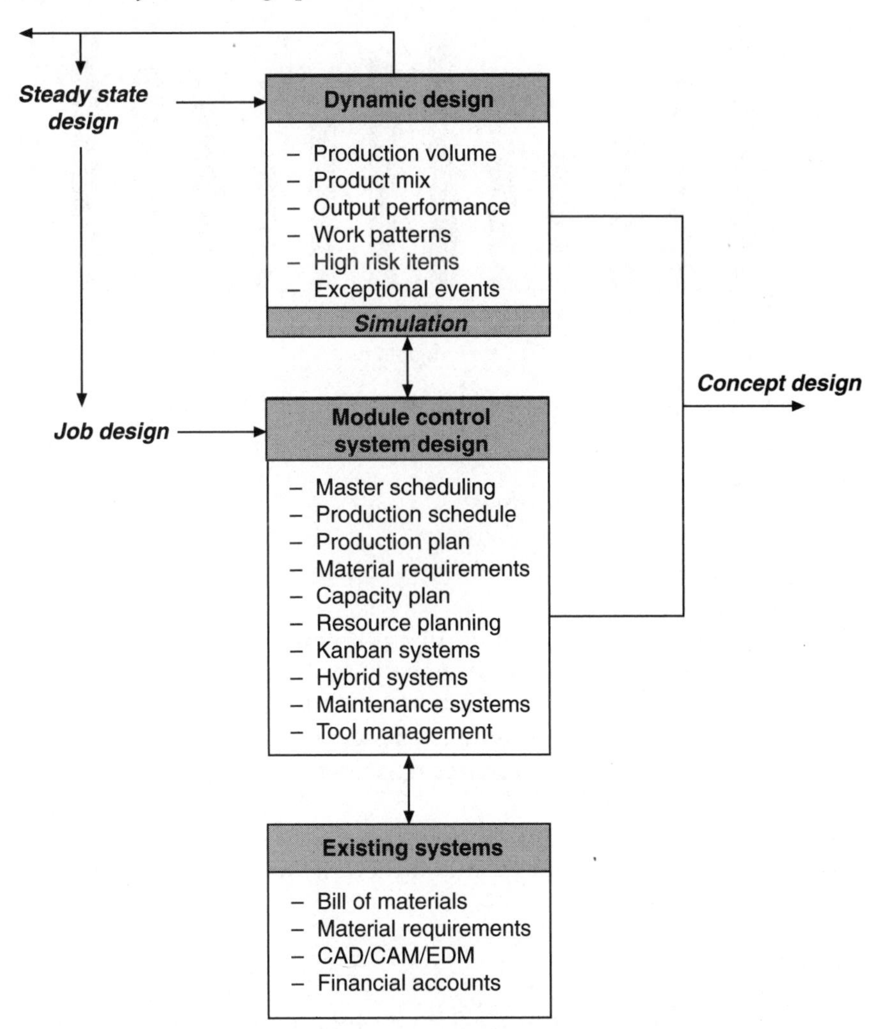

6000 units per month (as used in the steady state design), then alternatives must be examined, such as:

- Increase available capacity:
  - speed up process;
  - increase time the machine can be made available;
  - purchase additional capacity to cope with peak demand;
  - invest in extra equipment;
  - increase manning levels.

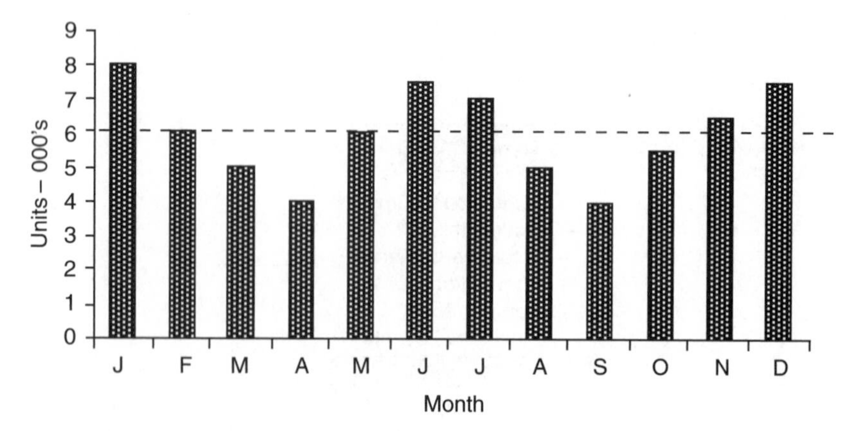

Figure 4.1 Example of demand variation for range of components

- Eliminate non-productive time from the process:
  - reduce changeover times;
  - improve tooling and fixtures;
  - pre-manufacture, performing non-critical operations using other equipment;
  - confirm process capability of equipment;
  - check the capability of measuring systems;
  - introduce preventive maintenance routines;
  - improve process removing items generating waste;
  - train team members in alternative ways of working.

The optimum solution is normally a combination of actions that involve improved methods, smoother factory schedules and possible investment in tooling or equipment.

Product mix

The actual mix of product required from the module is dictated by customer demand and changes experienced by the market for their products. The situation at component level is further complicated by similar items being sold to more than one customer; increased market share is often gained at the expense of other customers. Predicting short-term changes in product mix within the overall volume requirements is extremely difficult, but the manufacturing system design has to incorporate ways of accommodating these variations without impacting overall output capacity or customer service levels.

Techniques for handling changes in product mix include:

- Adjust available capacity to accommodate the possible changes of mix.

- Increase capacity to handle variety and production changeovers.
- Attack changeover times on machines, assembly equipment and test facilities.
- Introduce preset, ready-for-use dedicated tooling and fixtures.
- Identify tasks that can be performed off-line and work that can only be performed once equipment is available.
- Establish a sequence of product changeovers to minimize the work needed to change from one item to another.
- Limit the number of tools available to design engineers to those in the machine tool carousel, so the only changes are to software.
- Practise changing over processes, establish standard procedures and essential tools.
- Redesign the manufacturing process to allow increased variety and small lot production.
- Evaluate alternative equipment capable of handling the range of product types.
- Examine stocking policy and the use of buffer stocks to smooth variation.
- Increase the team skill base to ensure everyone has the capability to perform all operations.
- Reduce the lead times for products, allowing items to be made to firm order requirements and not against market forecast.
- Manufacture in small batches so actual production matches specific customer requirements.
- Design common components suitable for different applications.

Again, the practical solution will be a combination of factors depending on particular circumstances and customer requirements, but the issues must be addressed and modifications made to the steady state system design to accommodate inevitable changes in customer demand.

Customer delivery performance

Customer delivery performance in most instances is dependent upon the robustness of the supply-chain process. All processes experience unplanned events that disrupt the flow of work. Most can be identified in advance and plans developed to limit their impact upon customer deliveries. Common events that disrupt production are:

- Material shortages from suppliers or component cell-remedies include:
  - maintain buffer stock of critical items;
  - hold additional raw materials stocks, allowing supplementary manufacturing routes to be initiated;

- interrogate spares and aftermarket stores for surplus items;
- examine the bottleneck operations stopping the flow of work (usually goods inwards inspection, surface treatment areas or outside processing);
- ensure payments for previous transactions have been made;
- provide good short- and medium-term visibility of schedules;
- establish simple blanket orders for low value items;
- take small items off the manufacturing requirements plan and establish replenishment Kanbans with the supplier;
- make the supplier responsible for maintaining adequate stock levels based upon actual usage;
- establish quality-assured preferred suppliers who operate in partnership and deliver directly to production, bypassing goods inwards and inspection;
- purchase subcontract machining as complete parts, making the supplier responsible for delivering quality-assured components on time;
- kit small items into an assembly pack, making the supplier responsible for packing all items ready for use;
- establish regular communication with suppliers to ensure they understand the business drivers and end customer requirements;
- match the lot sizes to actual production quantities;
- obtain regular deliveries in smaller batches;
- find local suppliers who can respond quickly to changes in schedules;
- work with suppliers to reduce lead time from receipt of order to delivery;
- improve the quality of information and data accuracy of computer systems; and
- confirm the lead times on materials requirement plan matches actual requirements.

- Machine and equipment breakdowns:

  - refurbish equipment prior to installation into cell;
  - build machine stoppage allowance into production plan;
  - build maintenance time into production schedules;
  - establish daily equipment checks and first-level routine maintenance;
  - introduce simple daily disciplines to clean machines, tighten all bolts and repair any obvious faults;
  - establish a plant history file for each piece of equipment;
  - introduce planned maintenance routines;

  - check guards for suitability, effectiveness and ease of operation;
  - eliminate oil leaks from machines or equipment.
- Quality losses:
  - increase capacity required to allow for scrap, rework and rectification;
  - build adequate scrap allowance into production volumes;
  - confirm capability of processes against the tolerances required in the specification, correcting any deficiencies;
  - check capability of measuring systems, ensuring they are suitable for measuring critical parameters;
  - establish skill level of cell team members;
  - train people in quality procedures and measurement techniques;
  - train people in the necessary skills and core competencies;
  - train team members in problem-solving techniques and design of experiments;
  - introduce fail-proof devices to prevent process errors occurring;
  - examine method of transporting parts, checking for physical damage;
  - introduce meaningful quality procedures and corrective action routines;
  - collect information and review the main cause of defects;
  - introduce a corrective action programme to identify and remove the root cause of defects;
  - identify the critical features for components, establishing how they will be measured;
  - document critical features and maintain formal records on quality performance;
  - introduce statistical methods for insuring process capability;
  - establish a senior engineering management review board to resolve recurring quality problems;
  - verify the calibration of equipment and maintain records of test dates;
  - control gauges and measuring equipment-verifying calibration checks;
  - introduce meaningful measures of performance, providing information on achievements in meeting customer expectations; and
  - hold regular group meetings to address quality issues and customer concerns.

A combination of factors and initiatives will provide the optimum solution, but these need to be considered as part of the dynamic manufacturing system design, introducing changes to the original steady state concepts in order to accommodate transient events.

Working arrangements and shift patterns

Working arrangements within the factory have a significant impact upon the manning levels for cell teams and investment in capital equipment needed to manufacture products and meet production schedules. A structure for shift working should have been decided at the outset, but policies for managing the actual time people spend working need to be established. Allowances will have been made for holidays and absenteeism in the initial design phase, but these require further consideration to determine working arrangements for managing supply-chain operations. Items that should be addressed include:

- Fixed holiday arrangements, closing facilities and adjusting the number of working days available for production.
- Flexible holiday arrangements, arranging additional cover for periods when people are absent.
- Agreed scheduled workloads, allowing the team leave when the work is finished.
- Alternative shift patterns, increasing capacity on bottleneck operations.
- Employment of additional people on routine tasks, transferring them into teams when required to meet peak demand.
- Employ contractors to perform routine jobs, when needed.
- Employ, say, 20 per cent of workforce on short-term contracts from an employment agency bringing people in as required.
- Use overtime and weekend working to provide additional capacity or cover for absenteeism.
- Contract-out support services, for instance canteens, facilities management, security and transport, and so on, making contractors responsible for managing own people and maintaining service levels.
- Introduce flexitime, giving people freedom to organize the times they attend work.
- Record holidays and absenteeism, making them a visible performance criteria.
- Training time requirements and scheduling it into the team's workload.
- Paying people to undertake training and participate in improvement groups outside normal working hours.
- Policies on attending dentists, opticians, routine hospital appointments, and so on.
- Procedures for attending family events occurring in normal working hours.

- Formal channels of communication, building time into work schedules for regular stand-up meetings.
- Working time for continual improvement teams, including time to develop the innovative ideas needed to transform the manufacturing systems into a world class competitive process.

Final detailed working arrangements and techniques for accommodating variations in work load can be refined after the new manufacturing process has been launched, but it is important to identify and agree how the business proposes to manage people at this stage in the design process. The actual number of people required within each area will change as teams go through the necessary training programmes and initial start-up problems are resolved. Continuous improvement initiatives and increased effectiveness as teams progress down the learner curve pose a challenge of how to adjust team size without impacting team spirit and morale.

Management of high risk factors

These items will vary depending upon the type of business and activities being performed. Therefore, it is impossible to identify specific factors that present the greatest risk to not achieving production schedules.

The following list illustrates some items that should be evaluated:

- Age and condition of the facilities.
- Process capability of equipment to consistently manufacture to specification.
- Robustness of the technology used in the product or process.
- Loading on equipment needed to meet the production schedule.
- Running speed of bottleneck processes.
- Heat treatment and surface modification processes.
- Inadequate material handling systems causing unnecessary damage to components.
- Items on long delivery lead times (forgings, castings, for example).
- Bought-out parts that depend upon craft-based, variable processes.
- New products and associated processes being launched into production.
- Interface devices that convert electrical signals to mechanical, hydraulic actuation.
- Obsolete items, particularly electronic components.
- Non-preferred suppliers, if the company is a non-preferred customer.
- Suppliers that cannot meet the required production capacity.
- Small and medium sized suppliers that are under-funded.
- Suppliers that are over 60 per cent committed to a single company.

- Complex test equipment not developed for production.
- Inadequate production scheduling processes.
- Poor data accuracy on information technology systems.
- Shortage of skilled workers for key processes.
- Lack of time and resources to train people in new skills.
- Moving or restructuring factories before undertaking a systematic manufacturing process design.
- Inadequately documented procedures that fail to provide user-friendly work instructions for teams to follow.

It is important to identify and assess the potential risk for all critical factors that could jeopardize customer service levels. These must be addressed and solutions found at the design stage, well before the cell is operational. If an issue cannot be adequately resolved, an action plan is required to minimize its impact following implementation.

Significant events

If the manufacturing system has been designed and implemented professionally, taking full account of the views and opinions of the people that will operate the system, these events should only be genuine accidents and one-off significant emotional events. Obviously, management cannot always adopt popular policies, therefore they must consider how to manage exceptional circumstances needed to engender changes in attitude, convincing people to accept radical new ways of working. Business and manufacturing systems design often involves consolidating facilities or movement of work between sites. These also need to be considered as an integral part of the overall process.

Exceptional items could include:

- Closing facilities and transferring work.
- Site consolidation selecting skilled workers, making others redundant.
- Purchasing a competitor and relocating resources.
- Moving facilities to new location.
- Cleaning up a site and removing hazardous materials.
- Selling product lines.
- Eliminating restrictive practices and traditions.
- Handling a strike situation.
- Renegotiating contracts, conditions of service and payment.
- Dealing with a serious in-service product failure.
- Reportable incidents and accidents.
- Exceptional events/accidents:

  - floods;
  - electrical substation failure;
  - roof collapse;
  - major equipment failure;
  - fires.

- Core process not performing to specification.
- Exceptional weather conditions preventing people getting to work.
- Loss of key skills through people leaving the company or not relocating.
- Accident involving core team members.

Significant events in some instances are predictable and the chances of them occurring can be assessed as part of the dynamic design process. They tend to be related to particular circumstances and require senior management involvement to determine the policy and take tactical decisions.

The dynamic design stage leaves many issues unresolved. They need to be critically assessed, in order to shape the ultimate solution that must be developed once the supply-chain and manufacturing systems have been implemented. The team responsible for a process must be directly involved in refining operating procedures and introducing practical solutions. Therefore, world class manufacturing is a combination of meticulous system design supported by everyone's commitment to continuous improvement and elimination of waste.

Process failure mode and effects analysis (FMEA)

A useful technique for enhancing the robustness of a supply-chain and a manufacturing system's dynamic design is the process FMEA. This is an analytical method developed by the automotive and aerospace industries for evaluating the possible failure modes of products. This technique has been extended to identify possible failure modes in supply-chain and manufacturing processes, determining the impact they could have upon customer service. A failure simply implies that the system is *not* functioning in the way envisaged by the steady state design. Further evaluation is needed to make a more robust process. The main steps in conducting a process FMEA are shown in Chart 4.2.

The risk priority number (RPN) is established by multiplying together the scores for *severity*, *occurrence* and *detection* to give a composite figure that represents the level of risk particular circumstances, transient situations or unplanned events pose to customer satisfaction (*quality, on-time delivery,* and *price performance*).

Chart 4.2 Failure mode and effects analysis process

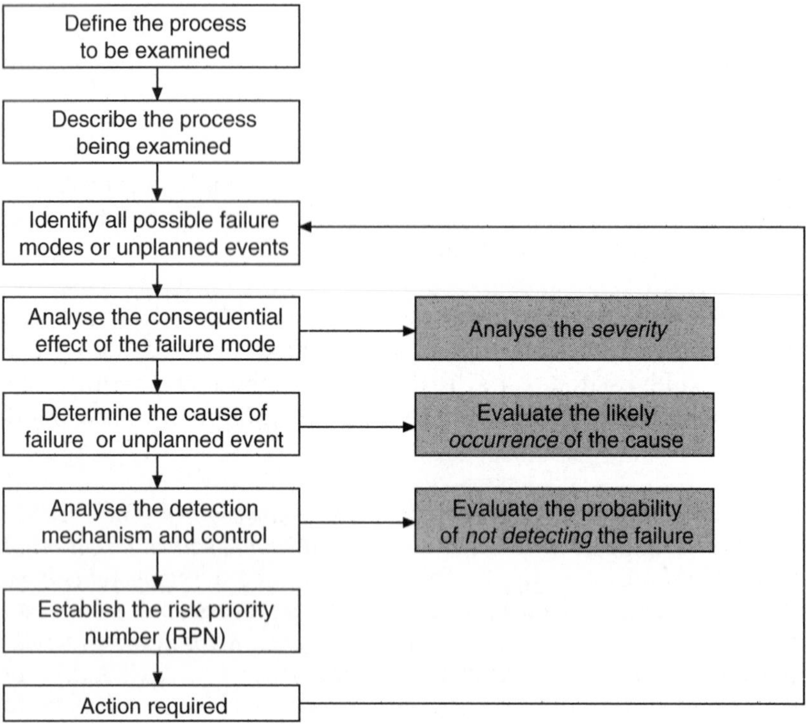

The *score* is determined using a predetermined set of tables calibrating the risk for each factor. A typical set of tables for supply-chain and manufacturing system design evaluation is shown in Tables 4.1, 4.2 and 4.3.

Table 4.1 Process FMEA for severity

Criteria	*Score*
Customer will probably not detect the failure	*1*
Customer will probably notice delivery performance degradation	*2*
Customer will notice decline in delivery performance	*3*
Factory reschedules production plan and informs customers	*4*
Customer will be subject to shortages but production not affected	*5*
Customer needs to reschedule production plan – able to recover	*6*
Customer reschedules production plan and informs their customers	*7*
Customer unable to meet delivery commitments to their customers	*8*
High degree of customer dissatisfaction	*9*
Potential safety problem caused by failure	*10*

Table 4.2 Process FMEA for occurrence

Criteria	*Score*	*Statistical proportion*
Remote probability	1	1/10 000
Distant probability	2	1/5000
Low probability	3	1/2000
Modest probability	4	1/1000
Moderate probability	5	1/500
Reasonable probability	6	1/200
High probability	7	1/100
Very high probability	8	1/50
Almost certain probability	9	1/20
Will occur	10	1/1

Table 4.3 Process FMEA for detection

Criteria	*Score*	*Probability %*
Remote likelihood event will impact customer	1	0–5
Distant likelihood event will impact customer	2	6–15
Low likelihood event will impact customer	3	16–25
Modest likelihood event will impact customer	4	26–35
Moderate likelihood event will impact customer	5	36–45
Reasonable likelihood event will impact customer	6	46–55
Distinct likelihood event will impact customer	7	56–65
High likelihood event will impact customer	8	66–75
Very high likelihood event will impact customer	9	76–85
Event will impact the customer	10	86–100

The FMEA analysis is undertaken by a team of experienced people who understand the weaknesses and possible failure modes inherent in the supply-chain and manufacturing system design. The team collectively identify all significant weaknesses in the process, making statements concerning:

- Definition of the process.
- Possible failure modes that could occur.
- Effect of the failure.
- Current method of control.

The team then considers each possible failure mode making informed assessments on relative scores for *severity*, *occurrence* and *detection* then calculating the *risk priority number* (RPN). This determines the relative degree of risk imposed by particular failure modes, allowing priority to be given to those posing the greatest threat to the business. A plan is then drawn up to *determine design modifications or corrective actions required for minimizing risk.*

This analysis is usually recorded in a standard format registering the scores and corrective actions needed to reduce the high risk factors to more acceptable levels (Table 4.4).

Table 4.4 Standard format for assessing possible process failure modes

Item No.	*Process*	*Failure mode*	*Effect of failure*	*Cause of failure*	*Current control*	*Assessment*				*Actions*
						Occ	*Sev*	*Det*	*RPN*	

The purpose of the process FMEA is to determine *What if ...?* – establishing ways of eliminating any possible occurrences posing a significant risk to operational performance or customer satisfaction. Recommended corrective actions should be evaluated using the same scoring system and process failure modes re-evaluated. This procedure should be maintained through implementation into production.

Evaluating system variables

Variability in manufacturing systems generally arises from two main areas, the availability of resources or the time taken to complete particular tasks. Both adversely impact performance making the system difficult to predict and control. The reasons for variation and ways of resolving situations are numerous, but for the purpose of evaluating system performance, only the different effects need to be quantified. For example, an analysis of machine stoppages may show the following contributing factors:

- Changeover times switching from one product to another.
- Adjustments and resetting the process ready for production.
- Routine and planned maintenance.
- Machine breakdown.
- Shortage of materials.
- Operator absenteeism.

Machining cycle times may be affected by:

- Setting errors.
- Poor process capability.
- Deficiencies in operator skills.
- Variable raw materials.
- Inadequate tooling.

For modelling purposes, the overall impact of these events has to be quantified into functional parameters that can describe the effect they have on the manufacturing process. The most comprehensive method for achieving this is to use an appropriate probability distribution curve, for example *normal, Poisson* or *negative exponential,* selecting the one that in practice most accurately fits the variable's behaviour. In order to simplify the model, attempts should be made to combine the effect of several sources of variability into a single distribution curve. For instance, it may be possible to describe all the influences on the cycle time by a single negative exponential distribution curve. Alternatively, some factors might have to be treated separately, with different curves being used for describing major and minor events. If it is not practical to perform such sophisticated analysis or the dynamic modelling system is unable to calculate the effects of applying distribution curves, it is still important to establish the range of values that are likely to occur in practice. This leads to an understanding of the probability of such events happening and provides information for performing sensitivity analysis on critical parameters.

Sometimes available information will not correspond to the proposed cellular manufacturing methods, so a comprehensive analysis may not be possible. However, information on the key processes should be available and experienced local opinion may provide useful sources for confirming basic assumptions. The most important factors to be determined are the demand rates and seasonal variations used to size manufacturing operations. All significant increases from current production levels must be rigorously challenged and confirmed in writing by the marketing team. Marketing forecasts are notoriously inaccurate and building excess capacity that is under-utilized, will destroy the technical credibility and commercial viability of the complete scheme.

Model building

Mathematical models are used throughout industry to predict performance. People apply simple mathematical formulae to describe the relationship between different parameters used to manage the business, but if oversimplified these models may not represent actual practice.

For example:

Output rate = Work in progress/Average lead times

Simply doubling the work in progress would not result in doubling the output rate; many other factors have an influence on the system. Therefore, accurate models must be derived using precise information. This usually involves calculating the state of each machine and operator activity within the cell at particular time intervals. This type of detailed calculation is very cumbersome to manipulate, requiring the use of computer spreadsheet packages or specialist software to support the analysis. Spreadsheet models cannot be used to assess the impact of random events, but can be quickly developed based upon the capacity calculations undertaken in the steady state design phase. These models are extremely useful for performing *What if ...?* and *sensitivity* evaluations because parameters are easily changed and corresponding calculations performed very rapidly.

Spreadsheet models have limitations in assessing a system's transient behaviour; in order to evaluate these a comprehensive simulation model capable of replicating the actual manufacturing system performance is required. Random events can be introduced and long operating periods running in accelerated 'real time' programmed into the model. More sophisticated packages can also show animated schematic pictures of the operation, aiding communication and explanation. Comprehensive models take considerable time to build and validate, but they can be used extensively to confirm capacity plans and test alternative operational scenarios, prior to implementation. The time needed to develop validated models can be several man-weeks, requiring specialist skills that may not be available within the system design team. However, the investment in building dynamic models can be well justified if the consequence of getting it wrong threatens future business viability.

Configurable simulators have the potential to simplify dynamic model building. These use pre-written systems tailored to particular types of problem, but care must be taken to provide data in the format specified by the software package. High level packages developed by specialist software houses are also available to make dynamic model building easier and more cost effective.

Sensitivity analysis

This determines how the system responds to critical parameter fluctuations, identifying key variables inside or outside the system having the greatest impact on business performance. The system's ability to tolerate these variables must be quantified and further design effort directed towards minimizing the impact they have upon operational performance. The model is tested using average values for the critical parameters, followed by applying extreme values for each parameter, assessing its relative impact upon operational performance. If changing parameter values shows little effect upon operational performance, then the variable can be considered insignificant. The evaluation should be continued using other parameters until important factors have been identified and degree of sensitivity clearly established. This technique is extremely useful, but it is difficult to test the interaction between two or more variables; in some situations insignificant variables applied together may have a dramatic effect upon the system.

Design refinement

The design can be refined by removing the variability source or modifying the supply-chain and manufacturing system to accommodate both uncertainties and the impact of transient parameters. The preferred approach is to eliminate or minimize the effect of variability before changing the system, but each factor must be considered in relation to its impact upon operational performance. Results from the sensitivity analysis and the process FMEA should have identified which factors are the main cause for concern, for example operator absenteeism, machine availability, material shortages through poor supplier delivery performance and such. These factors affecting the system have now to be studied in greater detail to identify the cause of variability. The easiest way to analyse this variability is to construct cause and effect (or fishbone) diagrams (Fig. 4.2).

A method for rectifying each factor can be identified together with the anticipated improvements. These can be used in the manufacturing system model to test the possible benefits and justify the cost of making necessary changes. The model should also identify when further development would mean only marginal improvements and when the analysis is complete.

Even at this stage some variables will still be considered outside the total control of the business. In these instances the manufacturing systems design must be made sufficiently flexible to cope with the fluctuations. An obvious example is variation in sales volumes; methods of accommodating such changes in demand have to be incorporated into the design. Ways of dealing with particular situations could include:

- Additional stock to buffer against seasonal demand.
- Extra machining capacity to handle transient overloads.
- Arrange subcontract capacity or people to cover additional shifts.
- Introduce flexible working arrangements maintaining production on bottleneck operations.

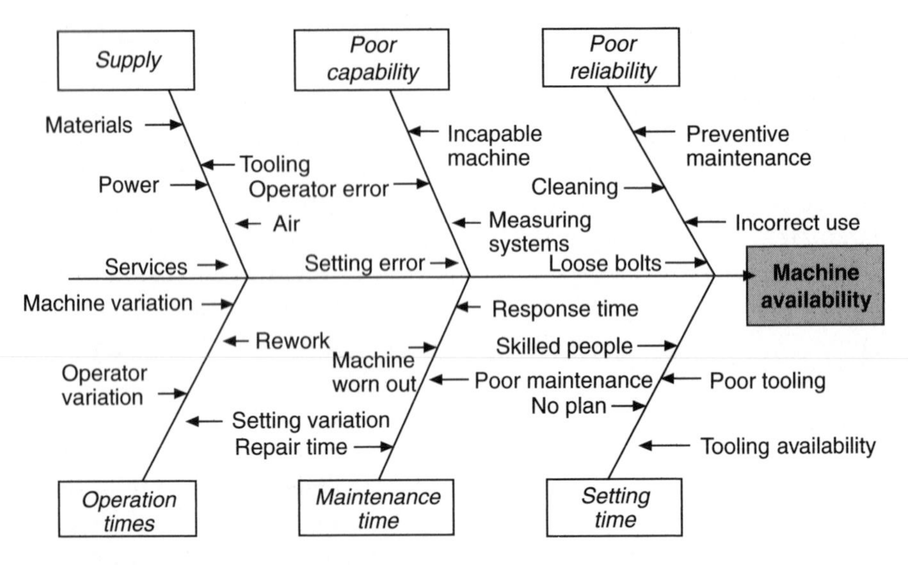

Figure 4.2 Typical cause and effect diagram for machine availability

However, the consequences of these revisions could have a significant impact on the initial estimates for the number of team members, machine requirements, investment in equipment, working capital, space, plant layout and such. Therefore, the feasibility of proposed changes should be verified, making necessary amendments to the steady state designs. Once the dynamic design is complete, the team should refine the supply-chain and manufacturing system specification, process diagrams, cell layouts, and so on, defining the performance envelope of the proposed production system. These summary documents should also state the accountability of the design team, the operational performance parameters expected from the system and the assumptions used for establishing the overall design. This specification should then be sealed and a formal process introduced to monitor and record changes, using a management review board to confirm if further revisions to the manufacturing process are justified.

Control systems design

The supply chain and in-house manufacturing processes have been described consistently as a 'system'. Processes need to be controlled like other physical systems in order to operate effectively. The job of designing integrated software-based manufacturing control systems is undertaken by information technology specialists; the task of the project team consists of compiling a detailed specification and rigorously selecting the most appropriate software package from a range of suitable suppliers. However, many companies will have already invested in manufacturing computer systems to be used across the company. These high level systems provide a method of consolidating information, storing it, linking it to other systems, and manipulating data into a format needed to run the business. If these all-embracing systems are used for controlling the business without simplifying processes, they are either too complex for maintaining meaningful information or too general, rendering the information practically useless for controlling actual processes. Therefore, the manufacturing design project team must understand the overall system characteristics, simplify the processes and integrate local control systems needed to manage day-to-day cell team activities. These must also reflect business policies and the desired ways of working believed necessary for enhancing business performance.

Master production scheduling (MPS)

The development of a realistic master production schedule is fundamental to controlling the supply chain and meeting customer commitments. It must be widely distributed across the business and suppliers as it:

- Defines products the business needs to manufacture.
- Establishes when products will be delivered to customers.
- Provides visibility of the future demand to suppliers, manufacturing cells, assembly and test areas.
- Determines the factory workload, both in-house and at suppliers.

Regular updating of the MPS establishes important disciplines for systematically evaluating and reviewing:

- Original equipment orders.
- Spare parts orders.
- Repairs orders.
- Future forecast demand.
- Arrears to customer orders and request dates.
- Provisioning for key items on long delivery lead times.
- Finished part stocks in the factory; assigned to customers, supporting future sales or obsolete.
- Product introduction and engineering development items.

The task of determining what manufacturing operations actually produce cannot be delegated. It is the sole responsibility of the most senior person on a site to own the master production scheduling process. This usually needs to be conducted at three levels:

Level 1 *Master production schedule* – orders, commitments or firm sales opportunities for future production volumes.
Aim: to determine the long-term capacity requirements of the production facility, examining the trends in demand over a six months to four year time span. Predicable changes in market demand are used to develop future business and site strategies.

Level 2 *Production schedule* – collates the production demand over the next six months or the lead time for procuring critical items.
Aim: to confirm the production capacity required, taking necessary actions to adjust the resource requirements and enable materials to be placed on order within supplier lead times.

Level 3 *Production plan* – establishes manufacturing commitment over the next reporting period.
Aim: to determine the actual mix of products the factory will manufacture including original equipment, spare parts and engineering prototypes. It should correlate closely with the agreed customer schedule making every endeavour to meet customer delivery commitments. It also attempts to smooth production demand providing level schedules and removing variations on the supply chain.

The master production scheduling process should be conducted using regular meetings with standard agendas aimed at minimizing the time needed to obtain necessary agreement.

The master production schedule at the highest level must be:

- Owned by the business general manager.
- Established by a team of accountable senior managers.
- Controlled and administered by a respected senior manager reporting to the supply-chain manager.
- Capable of providing long-term visibility of future requirements.
- Achievable within available manufacturing resources.
- Subject to regular review and governed by a set of standard procedures.

The master production schedule is translated into a production schedule and production plan identifying *what* the factory *will* manufacture and how long it will take to produce. Levels 2 and 3 scheduling processes must be owned by the site general manager, controlled and administered by a respected senior manager who is responsible for:

- Verifying that the sales value generated by the production plan meets the commitments made in the business plan.
- Confirming customer delivery requirements will be achieved.

- Ensuring factory loads match the resources available and the production plan can be manufactured on time.
- Attempting to smooth the demand profile, creating more even manufacturing schedules.
- Determining the assembly schedule that drives component procurement.
- Coordinating cell schedules and supplier commitments to delivering the production plan.
- Identifying critical shortages and problems that could impact customer deliveries.
- Owning the computer-based manufacturing system, taking responsibility for verifying and maintaining the data accuracy to a level of 98 per cent.
- Establishing the material requirements plan.
- Assigning categories to determine how parts will be treated within the material requirements plan.
- Identifying alternative procurement routes for obtaining components, balancing production loads when necessary.
- Establishing alternative and parallel production control systems for different categories of components.
- Monitoring achievements against the master production plan.
- Initiating corrective actions to recover missed delivery promises.
- Providing regular updates on measures of performance to pace the factory and demonstrate achievements.

The role of production planner is a key position and requires someone with an in-depth knowledge of the product range and methods of manufacture, or a bright graduate with extensive computer skills to manipulate the information.

These scheduling processes are fundamental to good manufacturing practice.

> *A supply chain cannot be expected to meet customer delivery promises unless it knows what to make and has the materials available.*

Material requirements plan (MRP)

The fundamental principle of such planning is to calculate component demand based upon the quantities identified in the production schedule. This calculation is relatively simple but depends upon an accurate *bill of materials* (BOM), realistic lead times for obtaining the components and confirmed production schedules. A product introduction management team should own the bill of materials with accurate lead times established by cell teams and commitments from suppliers.

However, a material requirements plan requires a more complex calculation:

New order cover = Total required – (number in stock + total on order)

This calculation must be repeated for designated time periods, determining net requirements for each future period. This is called *time phasing* with quantities being adjusted to reflect this phenomenon:

Net requirement = Gross requirement – (projected stock available + expected deliveries)

The full material requirements calculation combines time phasing and the dependent demand calculation for components at each level within the bill of materials. The planned orders are offset from their net requirements to reflect the purchasing and manufacturing lead times. The material requirements plan calculation is driven by lead times assigned to part numbers, but to reduce the complexity of entering data for each part, constant time factors are often used for families of components. This simplifies the process, but creates a major problem of how to maintain accurate data on the manufacturing database for each part with respective processing times across different work centres. However, every opportunity must be taken to refine information and reduce lead times on the computer system, because this governs when materials are ordered and released to production. Unless lead times are systematically corrected, any attempt to reduce levels of work in progress will be automatically thwarted by the computer system continuing to release orders based upon historic lead times.

Once a material requirements plan has been established the system generates:

- New production orders for manufactured and purchased items.
- Amendments to order quantities or due dates for existing items.
- Suspensions to existing orders or cancellations.

These calculations (for all but the most straightforward business applications) are very time consuming and if recalculated on a regular basis require support from specialized computer software. To prepare a complete material requirements plan takes considerable time even using a powerful computer system, but in most large manufacturing companies this is a high level application fundamental to managing a competitive business.

Refinements to the material requirements planning system

The first requirement is to reduce the number of levels in the bill of materials, simplifying the material requirements planning process. This is achieved by establishing a top level bill of materials for the assembly area and restructuring works orders for in-house component manufacture, making them compatible with the proposed module and cell structures. Only complete parts should be included, with any subassemblies either handled as 'phantoms' or ignored. This approach simplifies the manufacturing reporting structure considerably and reduces the number of entries required on the central computer system from entry/exit of each work centre to the entry/exit of each cell. It also means that local controls can be established for order quantities and batching rules more appropriate to the area. However, alternative cell-based scheduling techniques have to be introduced, assisting the team leader to plan and control work through the modules/cells.

A second way to simplify the material requirements plan and related purchasing systems is to collate small parts needed on particular products, defining them as a combined part number. These items can be purchased from a preferred supplier specializing in small auxiliary parts (nuts, bolts, screws, washers, seals, dust caps and such items). These are packed ready for use in assembly, supplied on time to meet the production plan and usually at a lower price than the cost of individual items.

The other technique worth considering is to place components into categories that determine how they will be treated. The number of categories required is dependent upon business complexity but examining components by Pareto analysis usually shows:

- 80 per cent of purchased items represent less than 20 per cent of the stock value.
- 20 per cent of purchased items account for over 80 per cent of the stock value.

The task is to assign each item on the material requirements plan a component category based upon the impact it would have upon working capital needed for running the business, while understanding that a shortage of any single item could stop customer deliveries and failure to meet sales targets.

These categories could be used to remove complexity from the system:

Category 1 The 60 per cent of purchased parts that represent less than 10 per cent of the stock value.
These should be removed from the material requirements plan and automatic order amendment system, and purchased

in predetermined quantities. The simplest method is to use the delivery lead time for setting up a simple three-container replacement system. One in use, another in reserve and the third at the supplier being replenished. An empty container then triggers delivery.

Category 2 The 35 per cent of parts that need detailed factory planning. These items should remain on the material requirements plan, using the system to release orders for raw materials, components and works orders needed to initiate manufacture.

Category 3 The 5 per cent most expensive purchased items, or critical parts procured from an unreliable sole supplier.
These have the greatest impact on the value of the stock and should remain on the material requirements plan but also be individually monitored to control on-time delivery, restricting the outflow of cash.

The material requirements planning system is dependent upon accurate data with a representative set of embedded, manufacturing assumptions as the basis for calculations. If data in the system is flawed, any instruction generated by the material requirements plan will be false, causing numerous production problems and rendering the system a liability. Data accuracy should be greater than 98 per cent, which takes considerable effort to achieve and maintain.

The difficulties associated with using the material requirements planning system to calculate and plan the release of works orders needed to achieve satisfactory due date delivery performance are even more complex, because the cell design and policies adopted for running production also have a significant impact upon manufacturing performance.

Therefore, the following need to be considered for creating a predictable environment:

- *Level schedules* – established by intelligent production planning, smoothing the workloads and creating a consistent rhythm for production.
- *Regular work patterns* – removing variability and sources of uncertainty, increasing the flow of work and reduction of lead times.
- *Manufacture in smaller transfer quantities or batch sizes* – smoothing the flow of materials, shortening lead times and reducing the value of work in progress.
- *Increase plant capacity* – allowing the cell to accommodate peak demand and time to introduce manufacturing process improvements.
- *Kanban techniques* – complementing the material requirements plan, simplifying manufacturing control systems.

The material requirements plan is a fundamental planning tool, used by product assembly modules and component cells where items are segmented at different levels on the bill of materials. Cell teams use the plan as the basis for local planning and control, establishing the sequence of events needed to support the production plan.

Capacity planning

Building a representative capacity planning tool is crucial for effective production planning and control, because it allows the capacity in assembly modules and component manufacturing cells to be determined based upon the product mix identified by the production schedule. The most accurate capacity planning tools use individual cell models, integrating them into an overall system, but even sophisticated tools still require a cell leader's experience and judgement to confirm the team can realistically meet production commitments. Manufacturing is a multi-variable, complex operation. For instance, to optimize the manufacturing sequence for 10 products flowing through 25 machining operations would take years of computer time to process. Therefore, usable capacity planning models must be simplified considerably but still retaining critical features governing performance. The type of model developed for dynamic analysis provides the foundation of a cell-based system, but this needs developing further to correlate the model with actual practice, making it more robust and user-friendly for shop-floor use.

Manufacturing capacity requirement is determined by:

- Volume and mix of products on the production schedule.
- Work content of products and processes needed to fulfil the production plan.
- The flexibility of equipment and processes to handle the product range.
- Skill of the cell team and its ability to perform tasks.
- Capability of equipment to consistently repeat operations.
- Changeover times to convert between product types.

Available capacity depends upon:

- Queue lengths at each operation.
- Loading on bottleneck processes.
- Availability and reliability of bottleneck operations.
- Product mix on the production schedule.
- Number of people assigned to the cell team.
- Overall workload and assignment of priorities.
- Enthusiasm and commitment of team.

The available capacity calculations for cells making relatively low varieties of components still requires support from some form of modelling tool. These could be based upon static calculations:

- Standard spreadsheet package, developed to model specific cell configurations and sequence of time-based events.
- Manual load versus time tabulations, accumulating the work content for key workstations using time-based process charts.

Or simulation using computer models:

- Manufacturing simulation package, programmed to represent cell operations and flow of work in accelerated real time.
- Specialist software packages, embedded in the manufacturing resource or business planning system. (These systems have similar limitations to the material requirement planning system; that is how to maintain accurate detailed information.)

The level of sophistication depends upon the business requirements. However, static methods can be very effective, with dynamic events being evaluated by repeated *'What if ... ?'* trials.

Advantages of cell-based systems:

- Relatively quick to establish and obtain meaningful information.
- Readily available data to operate the model.
- Simple to understand and use by the cell team.
- Results concise, easy to interpret and apply.
- Inexpensive PC-based system easy to install and operate.
- Forms part of the cell tool kit.
- Can be networked to other computer systems for downloading material requirements plans, production schedules and production plans.
- Allows business-orientated performance measures to be monitored and team achievements reported.

Simulation models are becoming more widely used as they develop into user-friendly, higher level software packages. Such tools allow the whole process to be modelled in an interactive environment and the flow of materials optimized for particular factors.

Advantages from using simulation models.

- Provides a facility to model random events.
- Presents a visual picture of events in accelerated real time.
- Models can be linked to form integrated simulations for the production process.

- Changes to models can be made quickly and in some instances on-line to evaluate the effect of particular events.
- Results are presented graphically and in a range of formats easy to interpret.

Once the initial investment has been made in purchasing software, creating a correlated model and training cell team members, simulation-based systems prove a powerful tool for evaluating cell capacity (Chart 4.3).

Chart 4.3 Factors influencing factory capacity

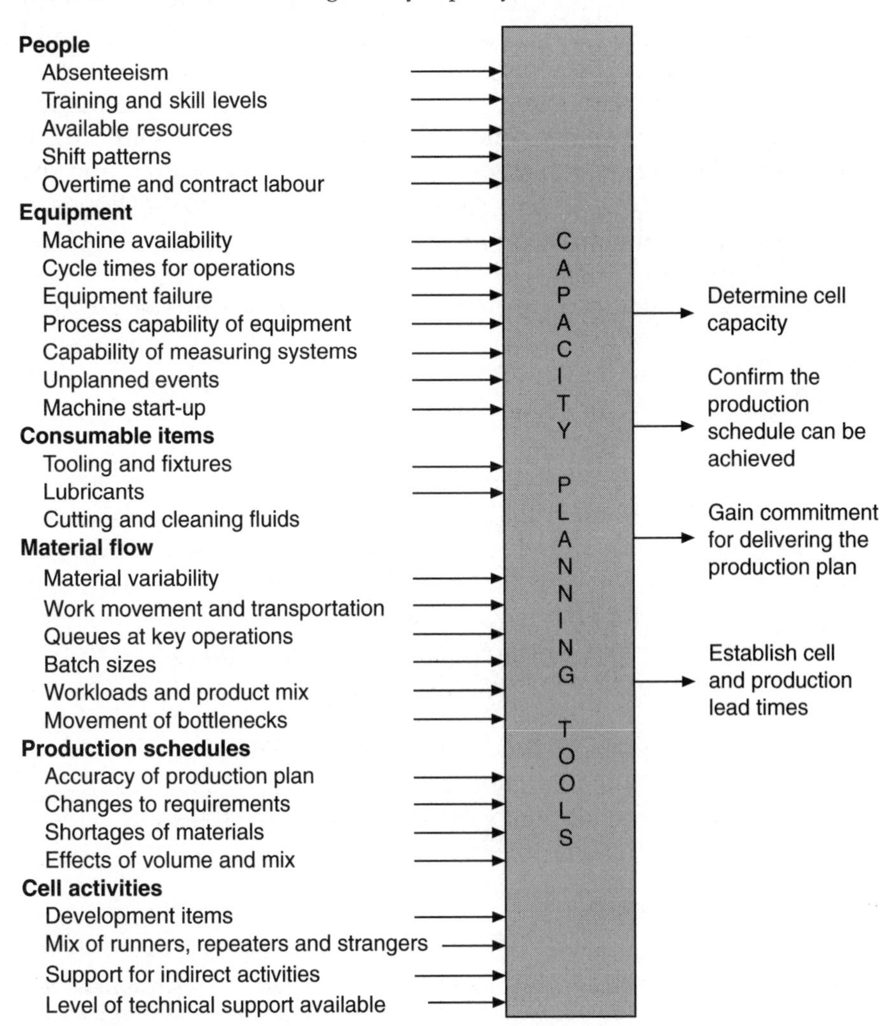

Manufacturing resource planning (MRP 2)

The material requirements planning process (MRP) is solely concerned with identifying materials requirements and does *not* include the labour and equipment needed for work tracking, capacity planning, requirements planning or controlling production (Chart 4.4).

Once material requirements planning systems were introduced on mainframe computers the functionality was enhanced by software suppliers seizing an opportunity to develop sophisticated packages that could control other aspects of a supply-chain process. These integrated manufacturing resource packages are often referred to as MRP 2 systems (Chart 4.5). They attempt to *'close the loop'* by taking account of the available capacity and capacity requirements plan to produce time-phased workload instructions for factory work centres. The systems consider new and current orders using finite and/or infinite planning routines to calculate the actual equipment and labour required for achieving the production plan. If the capacity planner cannot derive an acceptable solution, the production schedule or available resources have to be revised and the manufacturing resource planning software rerun. The calculations needed to produce representative time-phased workloads for each work centre are considerable, and systems often use net changes to requirements as a method of reducing computing time. These systems also provide business

Chart 4.4 Typical material requirements planning system (MRP)

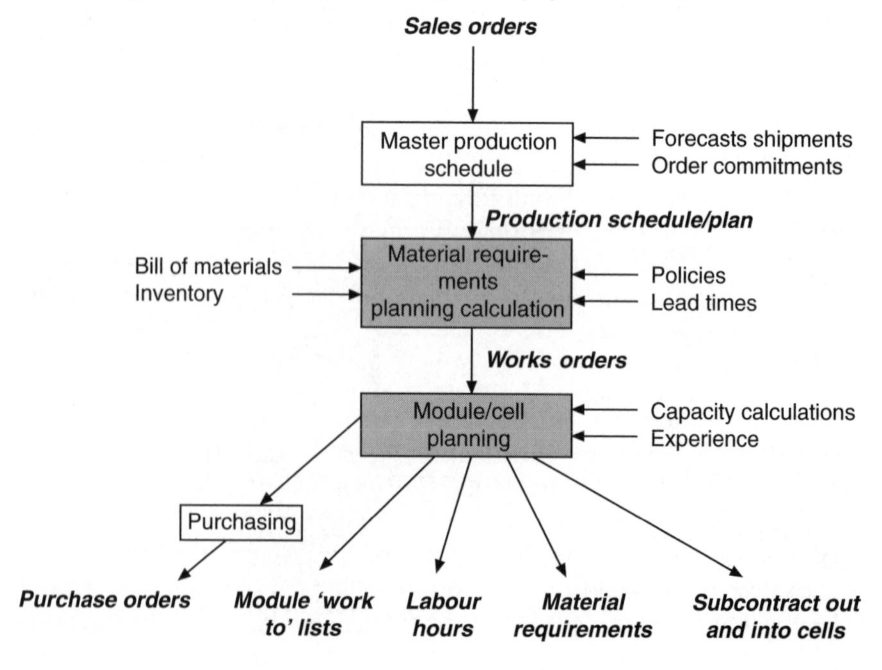

Chart 4.5 Typical manufacturing resource planning system (MRP 2)

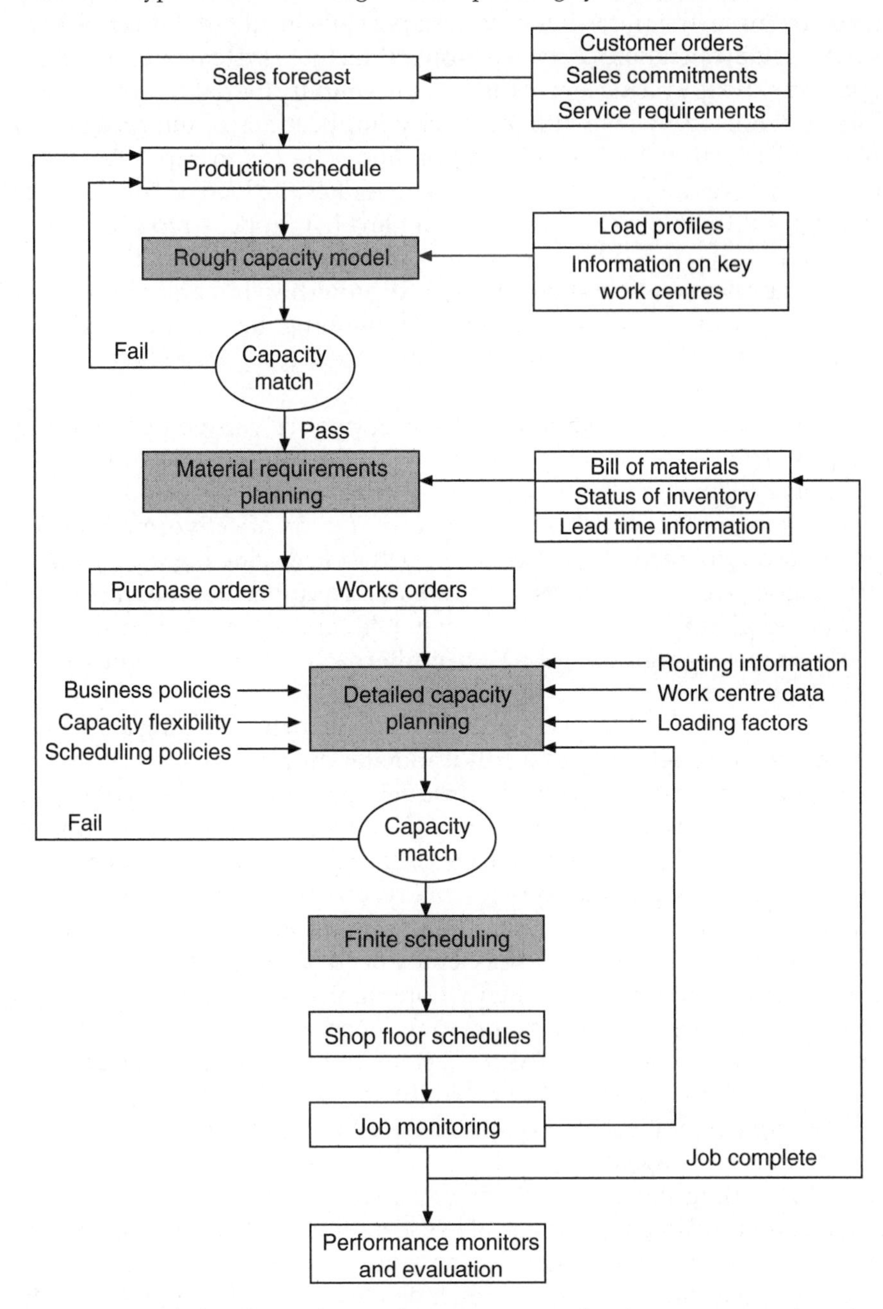

support modules for sales order entry, purchasing systems, invoice matching, work tracking and links with other financial, engineering and quality systems.

Experience has shown that to maintain accurate data at work centre level for these integrated systems is impossible in all but the simplest of supply-chain operations. Businesses operating these systems adopt general approximations to actual events, in order to simplify the data requirements. This is achieved by evaluating capacity implications of the production plan prior to running the full programme. The rough capacity model usually ignores all current orders and generates the load, using the total throughput time identified at a product level, as opposed to equipment level, and is restricted to considering only key work centres. Successful MRP 2 applications are those limited to providing integrated business support, performing high level activity planning, linked into local PC-based module/cell planning systems used for controlling actual events.

Modules and cells manufacturing a variety of components require support from local planning tools. The advances in microcomputer technology have stimulated the development of distributed manufacturing resource planning systems and are proving an effective method for planning module activities. They can be networked, allowing works orders to be passed electronically between cells and equipped with sophisticated cell-based planning tools to verify available capacity, marshal resources, track work in progress and identify alternative recovery routes from unplanned events.

The success of MRP 2 systems depends upon manufacturing managers having clear objectives and a full understanding of the tasks they are expecting it to execute. Simply introducing an all-pervasive computer system will probably not improve manufacturing effectiveness. A problem of using standard integrated packages is that they provide over-complex, difficult to maintain solutions for many relatively simple operations. Managers should identify where a simple approach would be more effective, where alternative processes are needed for handling particular situations, developing rules when a more rigorous procedure is required. In many instances, a business may only need a subset of the packages available in an integrated software system, but the prices charged by the vendor normally reflect its full capability.

The implementation of a computer-based integrated business system is a considerable task and again should be the responsibility of a dedicated team. Designing the manufacturing system should always precede the specification and implementation of new computer systems, as a significant part of the design task is introducing simplified processes with people taking responsibility for planning their own work. It is therefore advisable due to the availability of finite technical resources to complete the manufacturing process design and implementation project before embarking upon a major computer system installation.

Adam Smith in his *Wealth of Nations*, published in 1776, explained his experiences gained in a pin factory. He discovered productivity could be improved dramatically by breaking the manufacturing process down into a number of simple tasks, performed by a specialist. In this instance the process was simplified but the control systems needed to marshal work between operations that became more complex. This traditional philosophy has pervaded our thinking for the past two hundred years as people have endeavoured to keep processes simple at the expense of ever more complex control systems. Modular manufacturing systems reverse this basic philosophy, exploiting the higher skill levels and flexibility of team members to perform several processes. For example, team members integrate the process needed to manufacture pins, thereby simplifying the control system to provide information on how many pins are required to satisfy the customer demand, with the team confirming it has materials and capacity to make them on time.

Pull systems

The MRP 2 systems previously described are regarded as *push systems*, because material is pushed in at the front-end process. *Pull systems* are based upon planning customer requirements at the assembly module stage, using Kanban triggers to pull components through part manufacturing cells and from external suppliers into assembly. This technique was pioneered by the Japanese and has proved a cost-effective material control system suited to products in regular production and is fundamental to lean manufacturing concepts. However, Kanban material control systems cannot be introduced without effective supply-chain and manufacturing systems, applying the concepts identified in the steady state design process.

To operate effectively, Kanban material control systems need:

- Simple module/cell structures with clear boundaries and ownership of process.
- Supplier integration – suppliers working in partnership, delivering high quality parts on time.
- Capable manufacturing processes and measuring systems.
- Short changeover times on equipment.
- Reliable machines, assembly equipment and test rigs.
- Level schedules that provide visibility of production requirements and a consistent work flow.
- Everyone in the cell team committed to high quality standards, continual improvement and the elimination of waste.
- On-line error detection and failure prevention devices integral with the process.

- Meaningful information, measures of performance and a data accuracy of greater than 98 per cent.
- Multi-skilled teams with agreed training plans to share/broaden knowledge.
- People working in a flexible manner who feel proud to be part of the team.

Mechanics of a Kanban

A Kanban is a token or card that signals the need to produce components or to move material from one station to the next. Many variations have been developed since the original Toyota 'two-card' system; and some techniques use other mechanisms for providing the signal. One method growing in popularity is to adopt material storage containers as the Kanban signal.

Two-card system

Toyota pioneered this system, where calculated levels of material buffer stocks are held at feeder and consumer workstations in a number of Kanban containers. When finished parts are removed from the consumer area, a *conveyance* Kanban is released together with an empty container that is taken back to the feeder station. Stocks of completed parts at the feeder station are held in full containers together with a *production* Kanban. Parts are authorized for moving forward to the consumer area by exchanging the production and conveyance Kanbans. Releasing the production Kanban back into the feeder cell provides the signal to remanufacture a quantity of parts, thus completing the cycle. The production Kanban contains details of the part number, precise quantity, operation details and other essential information; the various parts are manufactured in the sequence the Kanbans are released back to the workstation. Using this technique, a variety of different parts can be controlled; the system ensures that variations in mix or volume are passed down the system. The number of Kanbans released controls the stocks held at each workstation and these limited stock deposits synchronize production throughout the factory.

One-card system

One-card systems use the *conveyance* Kanban only (initially proposed by Professor Schonberger) which is used to control the movements of materials between workstations but does not initiate production; this is authorized by the final build programme, using fixed lead times to allow availability of parts. This approach works effectively if the relationship between component demand and final assembly is predictable, that is when:

- Products are simple to manufacture with few common parts.
- The product mix and volumes are stable.

The advantage of a one-card system is simplicity of operation, but if the products become more complex or the volumes unpredictable, other techniques are needed.

One-card trigger
The two-card Kanban system satisfies two control objectives:

1 Triggers the movement of material only when needed.
2 Synchronizes production at interdependent workstations.

This can be achieved with a *single* card, provided that the buffer stock is held where it will be consumed. Once consumed the empty container and Kanban card with the component's details and required quantities are returned to the feeder cell to authorize replenishment. Once completed, the same Kanban triggers the transportation of the parts back to the consumer workstation. This requires parts to be held with more value added content at the consumer workstation; in some instances this may be undesirable.

Kanban squares
This technique is useful for controlling work between workstations, providing a regular flow of similar products. A square is introduced between each workstation with a defined buffer capacity, normally less than five parts. The feeder station cannot complete the next part until space is available in the Kanban square. This technique highlights the bottleneck operation through the build-up of stock, limits the work in progress and demonstrates which operations can be combined, eliminating waste from the process.

Sizing Kanbans and number of bins

The Kanban container size is determined by:

- Convenience of handling, particularly if the operator is expected to lift and transport the container manually.
- Quantity of items needed to provide sets of parts at assembly. The number of parts in the Kanban container is often based upon 40 because it can be factored into 20, 10, 8, 5, 4 and 2 compared to 50 which can only be divided into 25, 10, 5 and 2.

The number of bins required to cycle between two work centres is more difficult to establish. Fundamentally it is based upon the number of parts needed to maintain production while a container is being replenished. This is complex when all the process uncertainties are taken into account. Adjusting the number of operational Kanbans allows work in progress to be manipulated, weaknesses in the production process to be exposed and necessary corrective actions taken to improve the overall robustness of the supply chain.

The Toyota deterministic method of establishing the number of Kanbans is relatively simple, but this is following several years of refining supply-chain processes, removing variation and rigorously eliminating the impact of unplanned events on the process.

Using this method the quantity of Kanbans needed to maintain supply over the replenishment period is calculated by:

$$\text{Number of Kanbans} = \frac{\text{Demand} \times \text{Lead time} \times (1 + \text{Safety factor})}{\text{Kanban quantity}}$$

- Demand – demand per unit time for the product at a particular work centre.
- Lead time – total time for setting the machine, processing the components, transporting them to the next workstation, waiting time for the Kanban card to be returned (triggering the next run) and waiting time before the next process starts.
- Safety factor – based upon the experience of previous events, for example variability of lead time, demand changes and waiting time fluctuations due to changes in mix.

Hybrid systems

Hybrid systems aim to use the best elements of both material requirements planning and Kanban systems. Material requirements planning systems are good planning tools but are poor at controlling materials; Kanban systems offer a good method of control but are ineffective in planning manufacturing facilities. Hybrid MRP-Kanban systems are a further extension of cell-orientated material requirements planning, providing high level production schedules and procurement planning support, but using Kanbans to control the movement of materials through production areas and into assembly. This simplifies the manufacturing control system because the conventional bills of materials (containing several levels depending upon the complexity of the product build profile) can be reduced to a single level because intermediate operations are controlled by the Kanbans. However, conventional bills of materials and material requirements planning systems are still required to determine process routings and product costs, but manufacturing information is documented on the Kanban cards which must be updated regularly to reflect engineering changes and current production methods.

Simple planning tools

Visual techniques based upon planning charts or process cards are very effective for controlling work within cells. Time strips are prepared representing the length of time needed to complete tasks at particular workstations, including machine setup, product verification and routine preventive maintenance. These are placed in sequence against a time-based planning board to show the required and available capacities at each workstation, including the time when jobs will be completed. Similar planning boards can be used to plan work sequence for team members if processes are constrained by people and not machine capacity. These simple visible systems designed to meet the specific requirements of a module or cell prove extremely effective control mechanisms, focusing activities on those needing priority.

Summary

The type and complexity of the planning and control systems ultimately required for managing a business are dependent upon many factors which must be systematically analysed, taking a management decision on the appropriate systems needed to meet business requirements. In my experience, before attempting to implement complex manufacturing resource planning systems, it is always worth spending extra time simplifying requirements, exploiting every opportunity for experienced people to influence decisions. However, once an effective planning and control system has been implemented it is fundamental to factory operations and must be adopted across the business, with data rigorously maintained by everyone associated with supply-chain and manufacturing operations. The task of work tracking is being simplified by using barcodes attached to manufacturing documentation and containers used for transportation. These can be scanned electronically allowing data to be entered directly into the planning system, establishing accurate work tracking information and the manufacturing statistics needed for calculating accurate product costs, schedules adherence, lead times, levels of work in progress, actual manufacturing times, production volumes, quality costs, and so on.

Maintenance systems

Maintenance practices within a company significantly impact operational effectiveness of equipment and time available for performing added value work. Traditionally these practices have been reactive, fixing the machine or test rig when it has failed. Many companies are currently introducing preventive maintenance techniques, by building routine equipment inspection and planned maintenance time into the production schedules.

Advantages of these techniques include:

- Greater confidence that machines will be available when required by production.
- Reducing the number of unplanned events which adversely impact production schedules.
- More effective use of skilled maintenance personnel and resources.
- Routine maintenance activities which identify and rectify problems before they become significant.
- Avoiding premature deterioration of equipment and machines reducing the need for costly refurbishment/replacement.

The cost effectiveness of planned maintenance is dependent upon several activities being introduced into normal working practice, such as:

- Machines regularly cleaned, all the bolts tightened and obvious faults routinely corrected.
- Detailed records obtained on failure modes and deterioration in performance characteristics.
- Equipment made available by production at agreed regular time intervals to service machinery and support facilities.
- Maintenance service records continually updated and cross-referenced.
- Maintenance costs monitored preventing expenditure on unnecessary items.
- Regular training for maintenance teams, preventing poor workmanship from creating equipment failures.
- Rectification or replacements programme to bring equipment to the required state of process capability.

Machinery and equipment will break down on occasions as it is subject to everyday wear and tear, contamination or possibly corrosion. A maintenance system is required to reduce the frequency and restrict the impact of these occurrences, linked to a mechanism for quick completion of any necessary rectification, cost effectively.

An effective maintenance system requires the following attributes:

- A support team responsible for ensuring the availability of manufacturing facilities.
- Procedures for:
  - routine daily maintenance;
  - fault diagnosis;
  - breakdown maintenance;
  - planned maintenance;
  - preventive maintenance;
  - checking process capability of equipment; and
  - refurbishing equipment prior to installation.
- Identified training requirements for maintenance team and module/cell members.
- System for recording machine specifications and maintaining operating manuals.
- Policy for obtaining spare parts and identifying critical replacement items.
- Control procedures for the purchase, storage and distribution of spares.
- Method of collating and updating machine histories.
- Special tools for supporting maintenance activities, including computer data logging systems, laser alignment checking equipment, vibration analyser and such.
- Formal and informal methods of sharing information and operational experience with users of similar equipment.
- Direct influence on the purchase of new plant and equipment.

Measurement of maintenance performance

Performance measures are key drivers for improving the effectiveness of a maintenance system. Appropriate measures must be selected focusing upon factors critical to business performance; these could include the following (introduced at business, module or cell level depending upon the situation):

- Availability of critical machines.
- Overall plant and equipment availability.
- Total number of breakdowns.
- Response time to attend and rectify unplanned events.
- Time spent on planned, preventive and reactive work.
- Overall cost of maintenance:
  - lost production on critical machines;
  - management and technical support for maintenance activities;

- labour employed on site, under service contracts, and facilities management;
- spares inventory and maintenance stores;
- disposal of redundant equipment and reinstatement of facilities; and
- refurbishing machines.

Information should also be collected for key items of equipment, which directly impact customer service levels:

- Time lost from production due to routine, planned maintenance and unplanned events.
- Number of breakdowns and the mean time between failures.
- Average time to repair faults.
- Service intervals and maintenance requirements.
- Cost of repairs, spare parts, labour and technical supervision.
- Performance comparisons with similar equipment.
- Time to respond to breakdowns and identify the fault.

Maintenance systems have to be designed to meet business requirements; no single technique can provide an economic solution for the full range of equipment. Systems have to be devised providing cost-effective solutions based upon an integrated approach to equipment maintenance. This will require cell teams to be involved in first-line routine maintenance activities, keeping detailed records of events resulting in lost production. Dedicated maintenance teams perform preventive maintenance and rectify faults; third party specialists perform routine, planned servicing and repairs on critical plant and equipment.

The implementation of effective maintenance systems impacts directly upon:

- Increased productivity through equipment availability.
- Reduced cost of quality, lowering the number of rejects.
- Reduced overall maintenance and refurbishment costs.
- Lowered energy usage and increased efficiency of energy conversion.
- Reduced working capital through increased flow of materials.
- Improved health, safety and environmental standards.
- The working environment, creating a committed and caring business ethic.

Maintenance systems are a focus for continuous improvement groups, but it is vital to identify basic policies and maintenance requirements as an important part of the overall manufacturing system design process.

Tool management systems

Tooling in various forms is needed across the manufacturing process, on machine tools, in process areas, for assembly operations and product test facilities. In businesses that manufacture complex parts in low volume, planning and tooling control is as demanding as managing the flow of materials. Rigorous tool control is essential because tooling directly affects fundamental business drivers:

- *Process flexibility* – available tooling determines the component range that can be produced or tested on particular work centres; rapid changeover of tooling is critical for accommodating variations in production volumes and product mix.
- *Quality and process capability* – worn or damaged tooling directly affects product quality with increased production times.
- *Costs* – overall tooling and fixture costs far exceed the price of effective planning and control systems, and defective tooling or shortages invariably lead to production delays, wasted resources and lost machine capacity.
- *Lead times* – tooling deficiencies result in having to reschedule production or the purchase of special tooling, both causing increased lead times and possible failure to meet delivery commitments.
- *Changeover times* – the commercial viability of high variety, low volume manufacturing is governed by an ability to change over equipment rapidly from one product to the next while maintaining full process capability.

Appropriate tooling policies must be developed to address these issues. The importance of tooling has been recognized on automated machining centres with businesses funding the high initial costs needed for purchasing component mounting fixtures, tool holders, cutting tools and such. Without this investment (which can approach 30 per cent of the total machining centre costs), improved quality, productivity and efficiency benefits will not be realized. Conventional manufacturing and lean manufacturing principles are dependent upon similar consideration being given to overall tooling and fixture requirements, in order to make processes more effective by removing a primary source of variability.

Items that must be included in the overall *tooling package* are:

- Tool holders.
- Preset standard cutting tools.
- Die sets and specialist tools.
- Dedicated tooling for specific manufacturing processes.

- Fixtures for manipulating components, supporting intermediate operations.
- Manifolds and specialist items to increase equipment versatility or provide alternative process routes.
- Pallets for moving parts into and out of workstations and assembly fixtures.
- Dedicated clamping and mounting arrangements for performing specific tasks.
- Jigs to hold the workpieces at all stages of manufacture.
- Hand tools needed to set the equipment and make minor adjustments.
- Calibrated standard gauges and measuring equipment.
- Dedicated measuring systems.
- Other equipment:
  - changeover aids and lifting equipment;
  - cleaning materials;
  - pallets and boxes for protecting and storing parts;
  - trolleys for moving work;
  - items for routine equipment maintenance; and
  - lubricants.
- Consumable items, such as cleaning fluids, grinding wheels, coolants, filters, welding rods, specialist gases and such.

Items of tooling and fixtures fall into three main categories, according to how they are used:

- *Set-up tools* – needed before the job is processed but not during processing, so must be planned ahead of production and released once production has commenced.
- *Workpiece fixtures, tool holders and measuring systems* – needed to establish the production process and required throughout the operation.
- *Production tools* – perishable items used in the production process.

Cutting tools

Tooling has become a significant cost factor in manufacturing processes. It is often understated because factory accounting systems only record a tool's purchase price, excluding the direct costs of purchasing, storage and refurbishment. These accounting systems are also unable to identify the significant cost penalties associated with inadequate tooling such as lost production, increased machining time and poor quality, resulting in possible non-conforming parts or failure to meet customer delivery commitments.

The overall tooling costs can be evaluated by examining:

- Annual spend on purchasing standard tools and measuring systems.
- Annual spend on associated equipment:
  - tool holders;
  - pallets and boxes;
  - jigs and fixtures;
  - storage racks;
  - presetting equipment; and
  - calibration equipment.
- Annual expenditure on specialist cutting devices (hobs, diamond cutting wheels, etc.).
- Average value of tools, fixtures and gauges within the business.
- Annual turnover of tooling and associated equipment per year.
- Cost of raising orders with suppliers.
- Ratio of overall cost of tooling to cost of raising orders.
- Cost of procuring emergency tooling and equipment, including lost production on machines and any premium acquisition costs.
- Proportion of tooling orders regarded as an emergency.
- Annual number of hours lost in production due to tooling problems:
  - waiting for tooling;
  - scrap and rework caused by defective tools;
  - extended processing times through using compromised tools or measuring equipment;
  - additional machine changeovers;
  - increased changeover times;
  - waiting for first article inspection; and
  - setting machines and verifying tooling.
- Annual cost of:
  - tooling stores;
  - tracking tool life;
  - tool maintenance;
  - presetting inserts;
  - calibration of measuring systems; and
  - rectification of measuring systems.

The magnitude of these associated tooling costs and projected savings determines the level of investment that can be justified on an integrated tool management planning and control system.

Tool management principles

Consideration should be given to creating a tooling classification system based upon the durability and usage, as shown in Table 4.5.

Table 4.5 Tooling classification system

Tool classification	*Returnable*	*Issued for life*
Perishable	Returned to stores for refurbishment (drills, end mills, special tools and such)	Scrapped once useful tool life has expired (carbide inserts, grinding wheels and such)
Non-perishable	Returned to stores after use (micrometers, gauges, fixtures, and such)	Permanently issued to cell (tool holders, specialist equipment, and such)

This basic classification shows that different planning and control strategies are needed for each group of tools. Manufacturing resource planning techniques have been established for materials, but these are not usually suitable for managing tooling requirements. A comparison of different materials and associated tooling characteristics (Table 4.6) shows how the systems must address differing features.

The inability of manufacturing resource planning systems to adequately

Table 4.6 Differences between production material and tool planning systems

Characteristics	*Materials*	*Tooling*
Flow	Materials issued and processed sequentially into finished article	Tools issued, used, returned to stores, refurbished and reused
Volume	Volume of material issued per component constant	Volume of tools issued varies with material cutting characteristics and condition of tools
Capacity	Given volume of material yields consistent number of similar parts	Capacity of tools reduce with time and wear
Use	Same material used to produce different components	Same tool can be used on many different products and machines
Substitution	Material shortages seldom lead to use of alternative specifications	Tooling shortages often result in using substitutes which increase variability
Accessories	Material used as supplied	Tooling requires accessories – as tool holders, offset data, gauges and such

control tooling is another reason for adopting visible Kanban techniques to control materials within the factory, and to introduce an additional system for tool management and control.

The aim of a tool management system (Table 4.7) is to:

- Identify the correct tools needed on a particular work centre for a specific combination of components.
- Supply tools in the right condition and with sufficient life to complete the job.
- Provide the necessary information needed to use the tools (including tool offsets), cutting speeds, feed rates and such.
- Make tooling available at the right time to avoid delays in machine setting or running production items.
- Deliver tooling to the correct place for a particular work centre.
- Provide an economic tooling service that accounts for price of tooling, tool life, product quality and other associated significant cost factors.

Table 4.7 Initiatives to simplify tooling requirements

Tool rationalization	Establish the variety of tools needed to manufacture core products and remove all surplus items of tooling.
Tool policy	Determine an available standard tool range and establish a control mechanism to prevent proliferation of specialist tools, especially by new product introduction teams.
Tool standardization	Specify the number of tools available to manufacture the complete range of components produced on critical work centres, minimizing set-up times.
Tool selection	Introduce an appropriate set of standard tools linked to formal procedures for obtaining specific tools.
Tool requirements planning	Establish the type and volume of tooling required to meet projected production volumes.
Tool inventory management	Establish and maintain a tooling inventory and status database.
Tool management	Define different methods for controlling the various tooling classifications.
Tool library	Keep records of existing and preferred tooling to be used wherever possible to manufacture new products.
Tool costing	Develop a realistic tool costing and monitoring system.
Tool purchasing system	Instigate a purchasing procedure that combines the total site requirements, establishing consignment stocks from preferred suppliers of standard and production tooling.
Tool performance analysis	Collect and analyse tool performance information.
Tool information management	Manage all tooling-related information, including verification and calibration of measuring systems.

Allocation and control of tooling

One of the fundamental concepts of modular manufacturing is to install all the necessary resources within the cell. Regarding tooling, this ideal must be balanced against duplicating the cost of tooling packages, including equipment needed to maintain and refurbish them ready for use. This normally results in a compromise solution, where tools and equipment used solely by a particular module are located within the cell and are the cell team's direct responsibility. Other tools that are expensive and used by several areas are assigned to a central storage repair and replenishment facility that operates a factory-wide system, servicing a number of modules.

However, in cells dedicated to runners, the demand for tooling will be repetitive. Here tools and equipment should be allocated and managed by the cell team, even if similar tools are used in other cells. Special purpose tools are managed centrally to prevent the cell becoming cluttered with seldom used items. If a central tool management system is required, all tools are individually identified using a structured coding system giving similar numbers to families of tools. This identification must be permanently engraved upon the equipment; it provides a key parameter for the tool management database.

This database is used for collating relevant information on the following tooling features:

- Purchasing, establishing preferred suppliers and tracking supplier performance.
- Provisioning, ensuring tools are available to meet production requirements.
- Monitoring usage of particular tools and incurred costs.
- Stock control, location records and availability status.
- Refurbishing requirements and identified methods of servicing.
- Calibration records, recall dates and monitoring equipment condition.
- Available life based upon the actual usage against specified tool life.
- Tracing equipment in the factory and its status of production availability.
- Specification of tools and associated equipment:
  - cutting speeds;
  - expected life;
  - suitable applications; and
  - repeatability and accuracy.
- Process planning of production parts, based upon cost-effective standard tooling.
- Recording the capability index of tooling and measuring systems for specific applications.

Tool and equipment storage

The method of storing tools is critically important both within a central area and also production cells. It is vital to clean and refurbish all tooling, making it ready for use prior to placing it in stores. Storage areas should provide:

- A safe and secure repository for tooling and equipment, preventing it from being taken and used without consent or consideration of the consequential lost production.
- Protection for cutting edges, surfaces and other important tooling characteristics.
- Controlled, safe and appropriate climatic environment for measuring equipment and gauges.
- An effective location and retrieval system.
- Good housekeeping disciplines, documentation and records.

Refurbishment

If the tools are to be held in the cell, ideally they should be located close by the machine or in a refurbishment and presetting area. Tool refurbishment normally requires additional equipment for regrinding, sharpening, repairing and establishing tool offsets needed by numerical control systems.

A policy is required that identifies where such tasks will be undertaken for each family of tools:

- Production module or manufacturing cell.
- Central tool room facility.
- Specialist subcontract tooling manufacturer.

Tool management systems

Tool management systems have to be developed for specific applications, depending upon product complexity and the production methods used by the manufacturing process. No standard solutions exist but by analysing the following aspects, an assessment of the business requirements and features needed for a robust tool management system should emerge.

- Production tooling requirements in terms of product volumes and mix.
- Range of tooling for key production processes to meet production requirements.
- Importance of tooling and consequences of failure on business drivers.

- Priorities and control objectives for the tooling system.
- Tooling costs that can be technically justified and supported commercially.
- Coding and classification systems needed for range of tooling used.
- Opportunity to rationalize tooling requirements by key processes and cell locations.
- Tooling on-site owned by customers.
- Tooling held by suppliers and used off-site.
- Overall tooling requirements, grouped into different categories to determine cost drivers, alternative tooling methods and cost-effective tooling provision.
- Tooling requirements plan at different levels in the bills of materials.
- Suitable control systems for tooling categories, location and method of storage.
- Supplier base and methods for procuring tooling.
- Information management using IT support to handle the volume of data.
- Electronically coded tools as a basis for the tool management control system.

Tool management is a complex operation that must be considered at the dynamic design stage of the overall process. However, provided that tooling concepts have been firmly established and the module/cell requirements identified, tool management control systems can be further enhanced after the manufacturing system has been implemented. In my experience, the most effective approach is to rationalize tooling requirements, and introduce simplified systems appropriate to the area being served; computer-based systems may then deliver real performance improvements. IT systems available for tool management and control are becoming more user-friendly and cost effective with the introduction of distributed PC-based systems. In some companies with complex tooling requirements, the tool management system is developed further, integrating it with the business management system to provide a comprehensive control and reporting system aimed at limiting expenditure on tooling and all consumable items used by production.

Summary

The initial manufacturing system design is now complete. All the key factors should have been identified and systematically considered, providing robust supply-chain and manufacturing systems (Chart 4.6). Once implemented the design must be further enhanced through continual improvement to create a truly world class process. *Without this level of detailed analysis it is unlikely that an optimum manufacturing solution will ever be realized.*

Chart 4.6 Summary of factors to be considered during the dynamic design phase

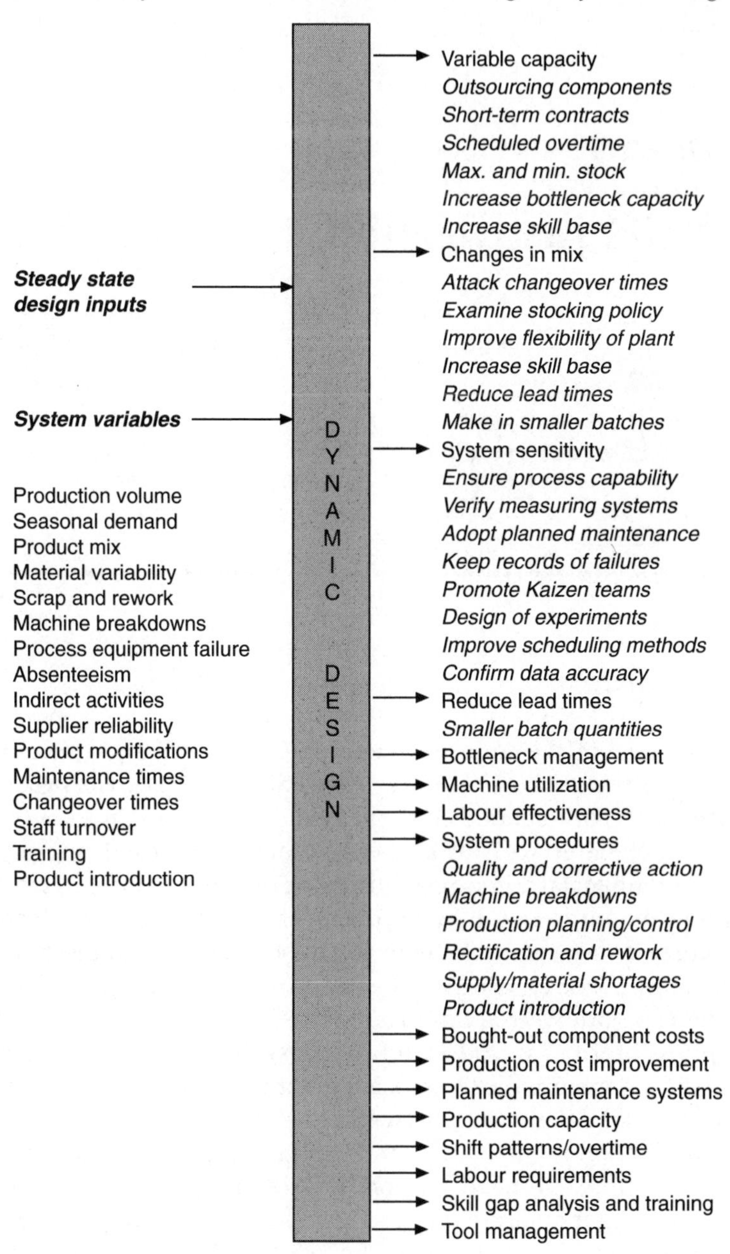

Chapter 5

Financial justification

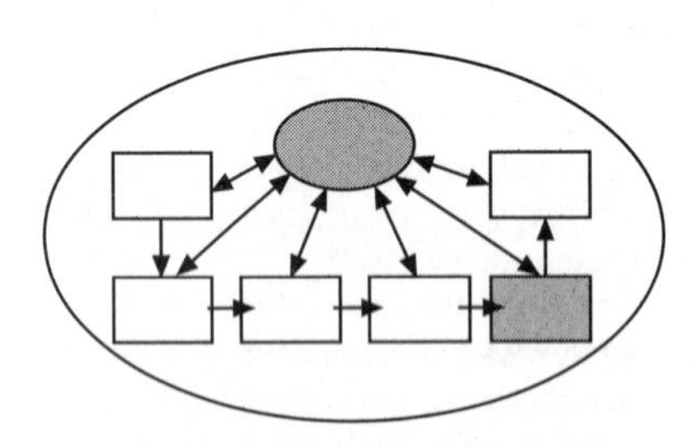

Financial case

The completion of the control system design effectively concludes the supply-chain and manufacturing systems design phase. The task now is to ensure that recommended changes can be justified financially through lower product manufacturing costs and that the proposed system meets business and financial criteria established at the project outset. Traditional accounting methods that may be used for calculating overall product manufacturing costs require refinement, allocating costs against particular activities as opposed to using overhead absorption techniques to generate a factory 'cost index'. It is important to establish accurate cost models that capture product introduction charges, manufacturing overheads, production and material costs associated with particular product families. They must also differentiate between the cost of 'runner' products in regular production from 'repeater and stranger' type products that require considerably more overhead support. It is important to prevent 'runners' from carrying additional cost burdens, which make them appear expensive compared to competitors' products. The cost of implementing the proposed supply-chain and manufacturing system (including the required capital expenditure) must also be verified against the commitments made in the business plan, with senior management support confirming that funds will be made available for the implementation plan (Chart 5.1).

Chart 5.1 Preparing the business case

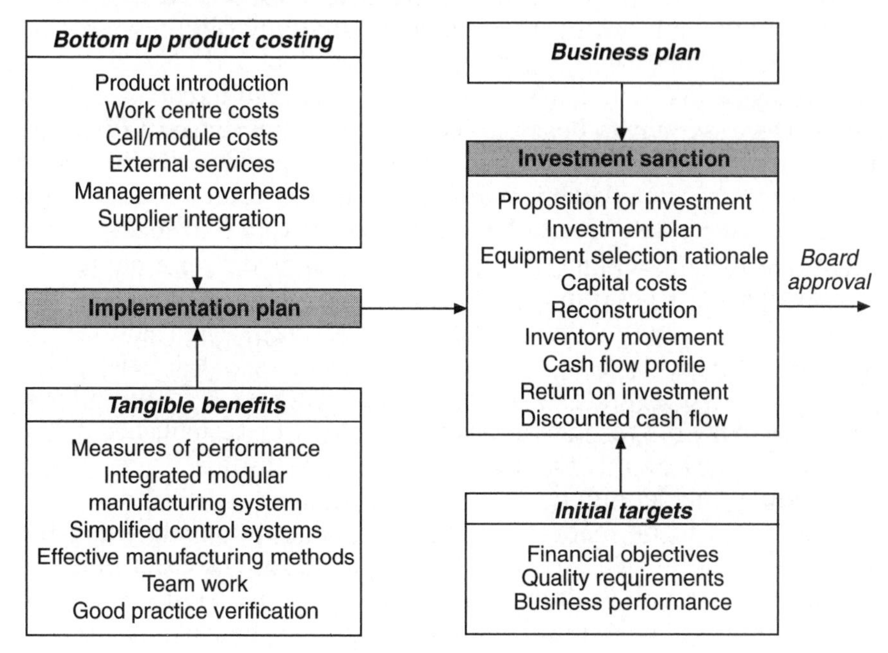

Activity-based product costing

Traditional cost accounting techniques measure resources used by production, but do not show *how or why* they were consumed. Since a variety of resources are required to perform different activities, the processes driving them are the product cost drivers. Once a business has been organized around key business processes and modules established dedicated to product families, then costs associated with each element of the process can be more readily identified. Costs can be allocated to specific product families in a direct relationship to the level of resources consumed.

The methodology for calculating base costs for particular elements of the production process is as follows:

- Identify the major process that the business performs and create a process plan linking the activities.

- Establish the cost drivers detailing the activities required at each stage of the manufacturing process.
- Cost the processes and activities (verifying that they add value, not simply cost) to create cost pools.
- Create a bill of activities for each product group.
- Allocate the cost pools in proportion to actual resources consumption for a particular product to create a representative cost model.

An activity-based costing system must be structured to collate costs at different levels within the business, because at management level a financial model cannot handle all the detailed information available for each element of resource associated with manufacturing particular product lines. Therefore, cost pools should be established at different levels in the organization, summarizing the resource requirements for a number of activities needed to perform various tasks. Cost pools for grouped activities become the drivers for the activity-based costing system and can be applied to product families creating representative cost estimates based upon the level of resources consumed. The next task is to apply these cost pools to the proposed module and cell structures established by the project team responsible for creating the manufacturing system design and module/cell definitions. This procedure allows realistic production cost estimates to be made for manufactured components using the proposed module/cell structure. The entire product costs must be determined based upon all the activities involved in manufacturing the family of product, including a proportion of management overhead costs, an element of product development costs, purchased items, facilities and such.

The previous items should be considered and included when calculating product costs.

Product introduction costs

Non-recurring product introduction costs for a family of products should be allocated based upon the expected product life using life cycle costing principles. The total development costs for a particular product line should be collated and recovered against the number of units to be manufactured each year over the expected product lifetime. The recovery rates can be adjusted annually to reflect the forecast production volumes and outstanding recoveries to be made over the remaining life projections.

Materials and bought-out components costs

These are based on the cost of items identified in the price file, but should contain a manufacturing allowance for waste, transportation, duty and packaging costs. Identifying these cost elements against specific products provides a more realistic figure. However, for the manufacturing system cost model other procurement-related overhead costs need to be considered. These include an element of supplies module costs, purchasing, goods inwards stores and quality assurance activities for validating items prior to use. Allocating these costs against the product families consuming the services significantly reduces the magnitude of arbitrary cost element, increasing the differentiation when considering product line profitability.

Employment costs
These should include employment costs plus a management allowance and an element of employment-related overheads such as cost of pensions, holiday pay, employer's national insurance contribution and medical insurance.

Work centre costs
The operating cost for each work centre within the cell should be calculated on an individual basis. These should include:

- Depreciation or operating leases for the equipment.
- Depreciation on capital needed for tooling, tool holders and fixtures.
- Replacement tools and fixtures.
- Maintenance and repair of equipment, including service contracts.
- Space rental charge proportional to the area occupied.
- Gas, electricity, water, fuel oil and such.
- Effluent treatment and disposal of waste materials.
- Consumable items.

Module and cell overhead costs
These charges capture the remaining costs incurred within the module/ cell that cannot be directly allocated against a product family. These costs include:

- Module supervision (although the cells should be managed by team leaders who spend the majority of their time performing direct activities).
- Information technology services, support and maintenance.
- Space rental for areas that cannot be charged directly to work centres.
- General tooling and consumable items.
- Depreciation or operating leases on module and cell facilities.
- Facilities maintenance charges.
- General maintenance and repairs.
- Remaining allocation of gas, electricity, water, fuel oil and such.

- Quality assurance costs for operating the system.
- Skills and team training.
- Continuous improvement groups.
- Fair allocation of manufacturing and production engineering costs.
- Transportation between work centres.
- Boxes and containers for moving work between work centres.
- Packaging and distribution.
- General management overheads.
- Other services used in the module, such as procurement, cleaning and security.

External services

The module design must strive to make all resources needed to manufacture a product available within it. In some instances, particularly for processes requiring expensive capital equipment, components may have to be sent outside the cell, sometimes to subcontractors. The cost of these operations can usually be directly charged to the product line receiving the service.

Holding costs of stock

The levels of work in progress and any necessary buffer stocks should have been identified as part of the supply-chain and manufacturing system design. The cost of holding stock at various stages throughout the process can be calculated and allocated to the product costs.

Internal quality costs

The cost of scrap and rectification including labour, material and work centre charges should be estimated based upon the process FMEA, manufacturing technical risks and known levels of process capability. This allowance for the cost of quality should be attributed to specific product lines.

Management overheads and selling expenses

These costs are normally outside the scope of the supply-chain and manufacturing systems design, but must still be allocated and used in product cost calculations. They can be unduly excessive in some instances, placing an unrealistic burden upon the product line cost base. These charges should be challenged and if necessary the business processes redesigned, making them more cost effective. The design of business processes is fully explained in *Plan to Win* (Garside 1998, Macmillan Business).

The product cost of particular products can then be calculated based upon the level of resources consumed, in relation to:

- Product introduction.
- Materials used for manufacture.
- Time people spend producing components.
- Resources consumed at each work centre.
- Bought-out components.
- External services.
- Time people spend on assembling and testing the product.
- Resources consumed on assembly and test equipment.
- Internal quality.
- Holding stock throughout the supply-chain process.
- Proportion of module/cell management charges.
- Proportion of management overhead and selling expenses.

The product cost for various items manufactured within a module must be financially modelled, using spreadsheets to apportion and collate the main cost pools. These should be critically reviewed and examined in considerable detail to determine the possibility of introducing less expensive processes, fewer resources or changed working arrangements to improve product line profitability. The manufacturing system design must also be critically reviewed to confirm an optimum process has been established, taking full account of the impact of cost drivers. If necessary, the module or cell structure must be modified to improve overall cost performance. These cost drivers also provide a focus for the parameters to be considered when selecting measures of performance, applied at business unit, module and cell levels.

Supplier integration

The trend in most manufacturing businesses is to purchase non-core items from outside suppliers, following a rigorous make versus buy analysis. This means that material and component suppliers control a significant cost element. A method of interfacing with suppliers must be established to consider total acquisition costs and to find ways of removing cost without impacting customer service. These changes impact upon both the supplier and the business as they work *together* to improve quality, delivery and price performance (Table 5.1).

Changes needed in working practices by traditional purchasing functions are numerous, transforming supplier relationships from 'them and us' to 'collaboration and partnership'. The aim of supplier development is to establish an environment of trust that promotes improvements in quality, delivery performance with a reduction in overall acquisition costs. This needs a similar transformation in thinking across the supplier

Table 5.1 Scope for supplier integration

	Possible current practice	*After supplier development*
Suppliers	Large number – multiple sources	Small number – single source
Sourcing policy	Price dominant factor	Quality, delivery and price
Stock	High material and component stock	Low levels of stock
Scheduling	Frequent change items expedited	Planned, communicated and level
Quality	High number of defects	Aim – zero defects, no inspection
Orders	Separate order for each lot	Long-term agreement, items called off as required
Delivery	Large infrequent batches delivered early or late	Small regular batches delivered just-in-time to meet schedule date
Payment	Late, taking extended credit	On time to the contractual terms
Product introduction	Late sourcing, following selection	Early involvement in the design
Systems	Inaccurate, complex, expensive	Collaboration verified using audits
Supplier assurance	Inspection of delivered items	On-site audits, corrective actions

base to that required when embarking upon the introduction of modular manufacturing systems (Table 5.2).

Working relationships between the business and the supplier must be jointly agreed with regular communication to understand how to meet purchasing requirements for operating in a modular manufacturing environment. Mutual benefits and cost saving can only be secured if both strive to remove non-value added activities from the overall process.

Table 5.2 Changes required to create an effective supplier base

Possible existing practices	*Requirements for modular manufacturing*
Check quality by inspection	Confirm process capability, build in quality
Long delivery lead times	Short delivery lead time against agreed plan
Infrequent deliveries	Frequent deliveries on a just-in-time basis
Large lot size	Small lot size, flexible processes
Poor performance, change supplier	Prevent poor performance by assistance
Constant renegotiation of price	Firm relationships with agreed cost reductions
Maintain distance in relationship	Make relationship an asset

Areas that must be considered include:

Production quantities

- Regular schedules placed on the supplier, whenever possible creating a steady flow of deliveries.
- Suppliers make frequent deliveries in small batch quantities, on time to meet committed orders.
- Delivery schedules and quantities committed to support the master production plan.
- Business to negotiate long-term contracts with key suppliers.
- Parts provisioned using blanket orders.
- Orders released using minimum paperwork and administration support.
- Contractual quantities agreed and confirmed.
- Actual delivery quantities variable, within the agreed terms of the contract.
- Total quantities of items delivered are as specified on the order.
- Suppliers pack items in exact quantities ready for use.
- Suppliers strive to reduce batch sizes, with assistance from the customer to help design the manufacturing system to operate more effectively.

Quality

- Suppliers establish own quality system, audited and approved by the customer.
- Customer provides assistance to supplier to implement necessary corrective actions, bringing quality standards to an acceptable level.
- Minimum number of customer specific requirements imposed on the supplier's quality system; however, most suppliers would be expected to conform to an appropriate ISO 9000 quality standard.
- Close relationships between buyers and supplier quality assurance groups.
- Suppliers strongly recommended to ensure full process capability of production equipment and measuring systems, implementing statistical process control as opposed to product inspection.

Suppliers

- Number of suppliers considerably reduced, most items 'single-sourced' with a key supplier.
- Repeat business automatically offered to same supplier.
- Standard items identified for specific products and purchased as a kit of parts from a single source.

- Active participation with key suppliers, designing the supply chain to remove non-value activities and improve competitive position.
- Competitive bidding limited to new part numbers and items which are continually inferior quality or subject to shortages.
- Manufacturing process for bespoke components identified in-house to establish cost-effective manufacturing methods and estimate target costs.
- Internal product cost targets and manufacturing methods used to identify suitable suppliers and support rigorous price negotiations.
- Price movement restricted, based upon established parameters agreed in the commercial contract.
- Suppliers encouraged and helped to design manufacturing process, introducing customer-focused modules to meet specific needs.

The benefits for the customer from this approach are manifest in the cost of parts, quality, administrative efficiency and delivery performance. The supplier benefits from continuity of orders, increased business, payment on time and increased confidence in the customer. The trend for businesses to outsource non-core components will continue, but as new technologies become more expensive to develop, decisions may have to be taken to acquire some core components from key suppliers or competitors. In these circumstances the need to establish partnerships and risk sharing agreements is even more important.

A key supplier working in partnership would be expected to:

- Provide the performance specification for their part of the system.
- Develop the technology to meet the specific requirements of the application.
- Design the interfaces to other subsystems in the product.
- Agree restrictive licences for the technology.
- Provide long-term supply agreements, preventing possible obsolescence.
- Give performance guarantees, including life warranties.
- Amortize development costs over the life of the product.
- Agree long-term price structures on original equipment items and replacement parts needed to support the aftermarket business.
- Support the development and performance qualification of the final product.
- Pay a proportion of the costs involved in verifying the product safe for public use.
- Enter into a risk and revenue sharing agreement, where appropriate.

The need for customers to encourage suppliers to be more cost effective and remain profitable is becoming increasingly important. The disruption caused by a supplier closing or being sold to another company can have a significant detrimental impact upon the business. Therefore it is the responsibility of the customer to ensure key suppliers are supported and helped to determine the most suitable manufacturing methods for their specialist components or services. The manufacturing redesign process can be daunting for small/medium sized enterprises (SMEs) to accomplish alone. Assistance from an experienced manufacturing engineer, seconded by their major customer to provide direction and support the business transformation, can lead to a dramatic improvement in business performance that must ultimately be of mutual benefit.

Purchasing organization

The structure of the purchasing function requires redesigning to meet the needs of a module-based manufacturing system. Traditional purchasing functions are unable to fulfil the broader role demanded by this approach. Therefore, a supplies module has to be designed which is capable of providing the appropriate service to both business and suppliers (Chart 5.2).

This usually results in a supplies module responsible for the activities shown in Chart 5.3.

Supply-chain support activities (including purchasing and procurement) are usually organized by commodity groups, which focus resources on areas requiring particular expertise. Their main concerns are ensuring parts will be available on time to maintain production and suppliers meet agreed performance targets on quality, delivery and cost. An additional purchasing business support group is also required, to be responsible for identifying supplier partnerships and negotiating keen acquisition terms on new product introduction programmes and longer-term issues that secure ongoing business profitability. These tasks require business-orientated, highly skilled teams, with members covering all aspects of professional purchasing.

The operational processes to be adopted by the supplies module need to be designed by a team, tasked with analysing the requirements of supply-chain modules, flow of materials, information, methods of communication and quality procedures needed to operate the system. They should adopt similar techniques to those used for designing manufacturing modules, making considered proposals on how each aspect of the supplies module should operate. These should be presented to the senior management steering group for comments and approval. The design team should make firm recommendations, providing a structure for the supplies module, job descriptions, outline operating procedures and a recommendation for staffing levels. This should be supported by

Chart 5.2 Process for designing a purchasing module

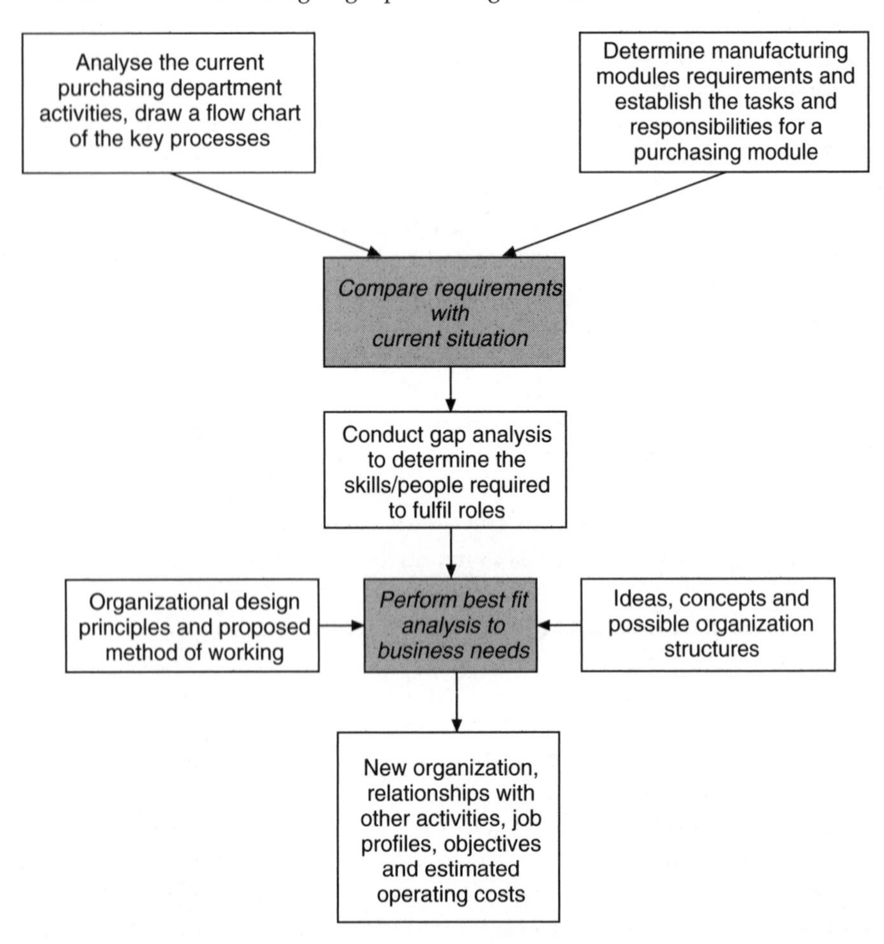

Chart 5.3 Typical organization structure and job roles for a supplies module

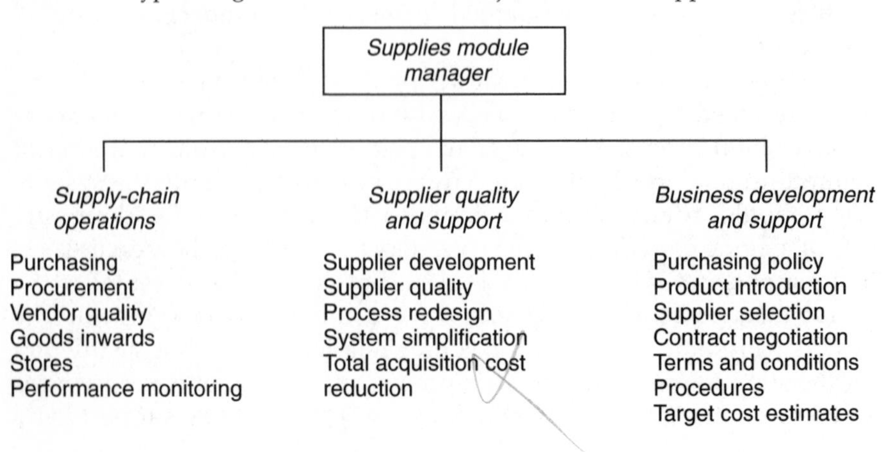

a critical assessment of estimated operating costs and a benefits statement, illustrating the performance improvements that could be expected following implementation.

Manufacturing measures of performance

Supply-chain and manufacturing performance measures are crucial for achieving the business's financial targets. Measures used at both the business level and within manufacturing cells should encapsulate the business objectives, focusing efforts on those factors that drive improved performance. Monitors have to be carefully constructed to allow collection of relevant information as part of a routine cell activity and not imposed thus creating a non-value added activity. The measures used at cell level need not be financially based, provided they give direct feedback on critical business parameters and cost drivers. The basic factors that have to be addressed at company, business unit, module and cell levels are:

- Adherence to customer schedule.
- Product quality.
- Stock utilization.
- Team productivity.
- Controllable costs.

The financial and non-financial measures that can be used for factory management information are broad, but each element must be precisely defined providing consistent data from different modules, or from a number of separate factories. The following items should be considered and the most appropriate parameters adopted across the business, provided that the information is needed to run the business effectively with the cost of the data collection not becoming a significant non-value added activity.

Labour/capacity

- Working days in month (or accounting period):
  - taken directly from the business calendar.
- Available hours in the period:
  - the hours the factory plans to be available for production – say 24 hours, 6 days per week.

- Total number of hours the factory worked in the period:
  - the factory is defined as working when, say, 80 per cent of the key machine tools and critical processes are manned and in operation.
- Total attendance hours – supply-chain operations:
  - total number of hours worked by module teams and people directly involved in supply-chain operations over the period.
- Overtime worked by cell teams:
  - number of overtime hours worked and payments made to personnel eligible for receiving paid overtime.
- Sickness absence:
  - number of hours sickness or absence by team members and people directly involved in supply-chain operations.
- Training time:
  - number of hours recorded as training for people in supply-chain operations.
- Industrial accidents:
  - number of accidents that result in one or more people being absent for more than three days.
- Environmental incidents:
  - number of spillages or releases with the potential to cause damage to air, land and/or water.

Ratios

- Percentage overtime/attendance hours.
- Percentage absenteeism/attendance hours.
- Percentage training/attendance hours.

Business unit and module productivity

- Number of employees assigned to module and cell teams.
- Number of employees required to supervise and provide general support.
- Area of floor occupied by the module/cells for equipment, work in progress and material storage.
- Number of units produced by the module team over the period.

- Number of standard hours or units of production produced in the period:
  - standard hours are a traditional accounting measure used in conjunction with overhead absorption techniques to produce a factory index rate allowing the rate of production and cost of manufactured parts to be calculated. (This standard hour measure distorts actual production costs and can drive factory effectiveness indicators in the wrong direction if applied too rigorously. For operations based upon high variety/low volume production of complex parts, then it may prove useful for comparing short-term performance achievements.)
- Sales value of production over the accounting period.

Ratios

- Sales per employee:
  - team members;
  - team members plus direct supply-chain support people.
- Added value per employee:
  - added value is more meaningful than sales per employee, giving an indication of the profit generated per employee.
- Sales to payroll costs:
  - monitors sales generated compared to employment costs.
- Added value to payroll costs:
  - monitors added value generated compared to employment costs.

Cost of quality

- Cost of failure:
  - burden on the business for not performing the work right first time.
- Internal quality costs:
  - cost of scrap and rectification including materials, labour, work centre charges, some management overheads and lost production capacity. (The cost of disruption caused by changing manufacturing instructions should also be included and charged to the product introduction team.)

- Quality assurance costs:
  - financial cost of operating a quality assurance system both in the factory and with suppliers. These should include the maintenance and refurbishment charges needed to ensure equipment is process capable, calibration of measuring systems, development of quality assurance systems, preparation of quality documentation/ procedures, including training and assessment by third party assessor.
- External quality costs:
  - warranty claims/cost of rectification, including product substitution programmes and third party settlements to retrofit products in service.

Ratios

- Quality impact upon profit percentage:
  - failure costs expressed as a percentage of trading profit.
- Quality impact upon sales percentage:
  - failure costs expressed as a percentage of sales.

Machine capability and capacity

- Number of key machines:
  - number of critical machines, including assembly stands, test rigs and measuring systems that fall into one of the following categories:
    - ◊ required by production for, say, over 70 per cent of the time when the factory is working;
    - ◊ a bottleneck process/machine;
    - ◊ one that is crucial to achieving the output irrespective of utilization.
- Process capability (Cp) of key machines:
  - process capability demonstrates the ability of equipment to repeat the process within specified tolerances. (The value of the process capability index Cp is a standard calculation made after measuring the dimensions of a key characteristic for a number of machined components.)
- Number of machines or equipment:

- planned to be checked in the period;
- machines or equipment actually checked in the period.

- Total number of machines or equipment verified as process capable:
  - number of machines or equipment confirmed as having a process capability index Cp greater than 1.33. (For rapid identification of those machines with an unacceptable level of process capability, pre-control techniques developed by Peter Shainin (Bhote) can be used very effectively.)
- Percentage of key machines that are known to be process capable.
- Number of critical processes on key machines controlled using statistical process control techniques.
- Percentage of capable processes:
  - measure identifies the percentage of processes controlled using statistical process control techniques to a Cpk greater than 1.33 and demonstrates the progress towards achieving full process capability where the key characteristics are consistently monitored within specified tolerances over defined time periods (see p. 251).
- Measuring system capability:
  - evaluates variations existing within measuring system and demonstrates that any variations are 'an order of magnitude' less than the variations in the process being measured. The evaluation of a measuring system's capability is a standard routine based upon using Gauge Repeatability and Reproducibility procedures (Gauge R&R) or Isoplot techniques. An Isoplot is a graphical technique developed by Peter Shainin (Bhote) which provides a quick assessment of the measuring system's capability.
- Number of measuring systems:
  - planned to be checked in the period;
  - actually checked in the period.
- Number of measuring systems verified as capable in the period:
  - confirmed as being accurate to 10 per cent of the allowable product variation using Gauge R&R or Isoplots.
- Number of measuring systems in the module used to measure key characteristics.
- Percentage of measuring systems known to be process capable:
  - monitor identifies the percentage of measuring systems used for key characteristics that are known to be accurate to less than 10 per cent of the allowable product variation.

Ratios

- Percentage planned downtime/available hours:
  - planned downtime on key machine tools, including test rigs and other key equipment as a percentage of available hours. (Downtime is when equipment is not operational within the available hours.)
- Percentage unplanned downtime/available hours:
  - unplanned downtime of key machine tools, including material shortages, as a percentage of available hours.
- Percentage efficiency:
  - time machines/test equipment performs useful work once the equipment is set up with tools and parts loaded, expressed as a percentage of the actual time the equipment is manned.
- Percentage utilization:
  - actual time machines are manned expressed as a percentage of the available hours in the period (available hours, say 24 hours per day, 6 days per week).
- Percentage unmanned running on key machines:
  - time machine or equipment runs unmanned expressed as a percentage of the total available hours. 'Unmanned running' means the machine is running productively without a person being assigned to operate the process, apart from loading or unloading the machine and checking process capability. This should occupy less than 20 per cent of the person's time.

(These measures should initially focus upon the bottleneck processes and key machines, moving onto other equipment once critical problems have been resolved.)

Stock

- Value of stocks:
  - total value of production items held as inventory.
- Value of the parts and replacement items for plant and equipment held in stores.
- Value of the tools, holders, fixtures and gauges held in stores.

Ratios

- Stock to sales ratio:
  - total value of inventory divided by the annualized sales in the period.
- Percentage stock held as raw materials:
  - value of raw material stock as a percentage of total inventory. (Raw materials are defined as items that require machining or processing operations, prior to assembly or shipment to the customer.)
- Percentage stock held as bought-out components:
  - value of the bought-out components as a percentage of the total inventory.
- Percentage work in progress:
  - value of work in progress as a percentage of total inventory.
- Percentage finished goods:
  - value of finished goods as a percentage of total inventory.
- Percentage stock held to support the aftermarket or replacement items:
  - value of items held specifically to service the aftermarket operation as a percentage of total inventory.
- Percentage stock held for repairs:
  - value of stock held to repair equipment in service.

Schedule adherence

- Sales value of the master production plan/planned sales in period:
  - ratio of sales value of line items loaded onto the production plan for delivery in the period to projected sales in the period.
- Percentage schedule achievement of the master production plan for the production period:
  - measures the business's achievement to deliver the master production plan including original equipment, aftermarket and spares. (Schedule adherence to the master production plan should be expressed as the number of line items delivered as a percentage of the total items committed. However, the correct quantities for line items must be delivered to qualify as adhering to schedule.)

- Percentage of the master production plan delivered in:
  - the first 20 per cent of the production period;
  - 40 per cent of the production period;
  - 60 per cent of the production period;
  - 80 per cent of the production period.
- Percentage schedule achievement of original equipment deliveries to the customer requirement date for the production period.
- Percentage schedule achievement of aftermarket and spares deliveries to customer requirement date for the production period:
  - schedule adherence to the customer requirement date should be expressed as the number of line items delivered to the agreed delivery schedule as a percentage of those items requested by the customer.
- Percentage data accuracy on the production control database:
  - data used by production control systems must have an accuracy greater than 98 per cent if the system is to provide meaningful information. To check this, a range of part numbers should be selected at random on the material requirements plan, counted and checked against figures held on the database. A statistical sampling plan should be used to confirm the overall accuracy level; any significant discrepancies must trigger a physical stock count.

Supplier performance

- Percentage of purchased items delivered to the master production plan requirements:
  - number of good quality line items delivered on time and in accordance with the agreed supplier schedule as a percentage of those items required to be delivered. (To qualify as being delivered on schedule the correct order quantities of parts must be received.)
- Items rejected from suppliers for inferior quality:
  - number of items rejected at inward inspection or in a cell prior to assembly, expressed in parts per million.
- Number of changes made to the customer schedule, following the agreement of the production plan, establishing production requirements over the planning period.
- Cost index:

  - cost movement monitored for a basket of items embracing 80 per cent of the cost of bought-out items for a typical product line. The price of these selected components should be updated continuously and movement for the reporting period and over the year to date expressed as an index against the previous year end figure. (Note, it is as important to monitor agreed cost reductions as it is to monitor cost increases.)
- Total number of suppliers – direct materials:
  - suppliers who provide materials and components used in the product.
- Total number of suppliers – indirect materials:
  - suppliers needed to provide materials, parts and services used by the manufacturing processes.
- Number of quality assured or preferred direct material suppliers:
  - who make deliveries directly into the factory without inspection.
- Percentage of suppliers who are quality assured:
  - number of quality assured suppliers expressed as a percentage of total number of direct material suppliers.
- Manufacturing lead time needed to purchase components:
  - elapse time between an order being placed on a supplier and items being delivered into the factory.
- Procurement lead time for delivering components:
  - elapse time needed for the supplier to fulfil the instruction to deliver parts into the factory.
- Frequency of deliveries made by key suppliers in the period.
- Number of supplier development engineers available to support the supplier base.
- Spend per supplies module team member:
  - value of items purchased in the reporting period per member.

Lead times

- Average system lead time for parts manufactured in-house:
  - average lead time loaded onto the manufacturing requirements

planning system for items manufactured in-house. (This parameter dictates when orders will be released for production and must be regularly updated or the system will automatically bring unnecessary materials into the factory.)

- Average system lead time for bought-out items and materials:
  - average lead time loaded onto manufacturing requirements planning system for bought-out items. (This triggers the release of orders on suppliers giving them authority to deliver.)
- Actual lead time for products supplied as original equipment:
  - time required for each family of products to be delivered to the customer once an order has been received.
- Actual lead time for current service parts:
  - average lead time required to ship current service parts after receiving a customer order.
- Actual lead time for non-current spares:
  - average lead time required by the business to ship non-current service parts after receiving a customer order.
- Average lead time for suppliers to investigate and deliver good quality parts, replacing defective items:
  - average lead time required by the supplier from the receipt of a returned item to instigating corrective actions, and deliver replacement items, excluding time spent waiting for a response sanctioning proposed changes.

Typical costs that can be allocated directly to modules

- Direct raw materials used to manufacture the product.
- Bought-out components.
- Labour cost of team members and people directly involved supporting manufacture – payroll costs including employment, benefit costs and paid overtime.
- Depreciation of equipment – charge made against capitalized plant and equipment installed in the cell.
- Space rental for the facilities – charge proportional to the area occupied, including rent, rates and facilities maintenance.
- Consumable materials – attributable expenditure on tooling, tool holders, measuring equipment, lubricants, solvents, cutting fluids, containers, boxes, cleaning materials and paper.

- Subcontract operations and processes–cost of purchasing additional processing that cannot be performed in the module, such as heat treatment, surface modification and specialist machining operations.
- Cost of gas, electricity, fuel oil and water–allocation made following fair assessment of usage.
- Maintenance and repair of equipment–charges incurred in repairing equipment in the module, using central services or outside contractors.
- Operating leases on equipment–rental premiums paid for equipment acquired on lease contracts.
- Cost of holding work in progress stocks in the module–value of work in progress. (This may be based on raw material costs or an increased valuation due to the type of work performed within the module.)
- Cost of scrap and rectification – cost of labour, materials and cell overheads needed to evaluate and remanufacture or rectify defective parts.
- General overheads and business management – costs associated with managing the business that can be reasonably attributed to the module, such as a proportion of the supplies module and purchasing expenses, central tool stores and senior supply-chain management.
- Other significant services required by operations – expenditure on essential services that must be contracted to specialist suppliers, such as third party quality accreditation, servicing and calibration of complex equipment, disposal of effluents and waste materials, IT and computing, transport, travel and communications.

An example of cell-based measures

Determining which parameters to monitor to capture essential management, financial and module operational performance information varies from being a straightforward to an almost impossible assignment, depending upon the complexity of the business.

A factory making diesel engine injectors equipment adopted what came to be known as *'Garside's MOPs'*. In this example, the method of obtaining module-based information was relatively easy:

- *Adherence to customer schedule* – check the number of items delivered each week against the production schedule. Only complete line item order quantities were acknowledged as meeting the customer requirements and the measure was obtained by calculating the ratio of line items delivered to the number on the schedule.
- *Product quality* – all rejects in the module were placed in a red container after the inspection reports and corrective actions had been identified. The weekly quality figure was calculated by counting the rejected

injectors, compared to the total produced and the quality performance expressed in parts per million.
- *Stock utilization within the cell* – at the end of the week the components at each workstation were counted. This was compared to the total number of injectors produced that week in order to calculate the module stock to sales ratio.
- *Team member productivity* – established by comparing the number of injectors produced to the number of hours worked by team members.
- *Costs directly controlled by the team* – monitored significant cost items that could be influenced and controlled by the team, such as:

  - overtime payments;
  - consumable tooling and solvents;
  - services, including maintenance; and
  - cost of quality.

In this instance the module and cell measures were easy to instigate, being directly relevant to the business objectives and running the module. However, considerable thought is required to identify which key factors could be used within the module to provide feedback on team member's performance and the essential management information needed to run the business.

It is important to inform management at business, module and cell levels of the achievements and contribution team members have made to improve business performance over a sustained period, and to say *thank you, well done!* Remember: *items that get measured, get done*!

Planning implementation

The first stage is to review the *good practice checklist* prepared at the manufacturing module and cell definition stage (see p. 64) to confirm all issues identified as important have been adequately addressed. If any significant factors have been omitted, the manufacturing system should be reviewed, making necessary changes to the process or working practices.

A plan for introducing a new supply-chain and manufacturing system is normally structured into three phases:

- Preparation (Table 5.3).
- Installation (Table 5.4).
- Implementation (Table 5.5).

Each phase has to be planned in considerable detail, using good project management disciplines supported by time-phased project plans, resource plans and clearly identified key milestones to monitor progress, see *Plan to Win* (Garside, 1998). Each phase has to be subdivided into inter-related work packages with time-phased integrated plans.

Work packages should specifically identify:

- The date when each task will start.
- Who is responsible for delivering it.
- People and resources needed to complete the work packages.
- Revenue expenditure needed to accomplish the various tasks.
- Capital expenditure required and dates when cash will be needed.
- Completion dates for key milestones.
- Budgets.
- Risk assessment for achieving deliverables and completion dates.

Table 5.3 Work package considerations for phase 1 – preparation

No.	*Work packages* *Phase 1 – preparation*	*Person responsible*	*Start date*	*Finish date*	*Budget*
1	Identify and select core implementation team, including project manager				
2	Designate a team office/chart room in factory				
3	Prepare board level paper for investment sanction				
4	Confirm performance improvement in business plan, restating projected financial performance				
5	Make detailed presentation with recommendations to senior management on proposed changes, including the expected costs and benefits				
6	Obtain formal senior management approval supporting the manufacturing system design concept and necessary capital and revenue funding				
7	Prepare detailed layout drawings for modules and cells, showing equipment position, including ancillary facilities needed within the cells				
8	Confirm operating parameters and specifications for capital equipment, including list of suitable suppliers				
9	Select equipment suppliers and obtain firm quotations and delivery commitments				
10	Determine how existing plant and equipment for new cells will be refurbished, ensuring process capability				
11	Confirm what refurbishment is needed for the building and floor areas of the modules/cells				

(Contd)

Table 5.3 *(Contd)*

No.	*Work packages* *Phase 1 – preparation*	*Person responsible*	*Start date*	*Finish date*	*Budget*
12	Prepare contractor's drawings for services, foundations and communication networks				
13	Confirm the work needed to reinstate vacated areas and environmentally clean up site				
14	Identify contractors and obtain firm quotations for services, refurbishment and restoration work with time scales and availability of resources				
15	Establish additional equipment needed to make cell operational – tooling, tool holders, fixtures, measuring systems, storage and transportation				
16	Identify containers and handling system for transporting materials and components within module and between cells				
17	Determine and specify specialist requirements for controlling the environment to meet cleanliness standards and customer specifications				
18	Obtain quotations for specialist equipment/ installation of environmental facilities				
19	Establish the information technology (IT) system requirements, links to other systems and training				
20	Establish costs for IT installation, hardware, software licences and system support				
21	Prepare outline descriptions for job roles needed in manufacturing teams and support activities				
22	Confirm number of people required to operate the new manufacturing system including support activities and supervision				
23	Agree selection criteria for appointing people				
24	Determine training requirements needed to broaden skills				
25	Identify a payment and reward system reflecting the new working practices				
26	Confirm the number and timing of people being released, transferred to other jobs or made redundant from the company				
27	Prepare the redundancy terms and support packages for those released				
28	Estimate the cost of releasing people and identify possible grants that may be available to enhance severance terms				
29	Prepare communications plans and hold discussions with representative groups to explain proposed actions				
30	Prepare quality plan and outline operating procedures to be implemented in the cell teams				

(Contd)

Table 5.3 *(Contd)*

No.	*Work packages Phase 1 – preparation*	*Person responsible*	*Start date*	*Finish date*	*Budget*
31	Assess training requirements for implementing quality system and procedures				
32	Identify who will deliver the quality and skills training, with time scales and costs involved				
33	Establish alternative ways of maintaining deliveries to customers while installing equipment, e.g. build contingency stocks before removing the existing process, find alternative routings, or purchase items from outside supplier				
34	Perform full risk analysis and prepare contingency plans to protect customer deliveries				
35	Develop communication plan for workforce, customers and suppliers				
36	Confirm installation requirements with phased module and cell launch dates				
37	Specify environmental, health and safety requirements for facilities and equipment				
38	Verify implementation and capital costs against budgeted figures included in business plan and subsequent investment sanction				
39	Confirm the specification for plant and equipment				
40	Conduct rigorous price negotiation placing purchase orders with preferred suppliers for new plant and equipment, refurbishing existing machines and installing facilities				

This first phase requires considerable management involvement and support to ensure that the wide range of tasks are completed and formally approved within a reasonable time. Many of the work packages can be undertaken in parallel, provided resources are made available to complete the work. If additional management resources to complement the core project team's efforts are not forthcoming, time scales for this preparation phase may become protracted or important issues neglected, placing the project in jeopardy.

Table 5.4 Work packages for phase 2 – installation

No.	*Work packages Phase 2 – installation*	*Person responsible*	*Start date*	*Finish date*	*Budget*
1	Confirm customer deliveries have been protected, making any necessary arrangements to retain existing critical processes for as long as feasible				

(Contd)

Table 5.4 *(Contd)*

No	*Work packages* *Phase 2 – installation*	*Person responsible*	*Start date*	*Finish date*	*Budget*
2	Remove available machinery to be installed in new modules and cells				
3	Refurbish key equipment to an accepted level of process capability				
4	Repair and paint any existing ancillary equipment transferred from other parts of the factory				
5	Clear and reinstate area in factory where new modules/cells are to be installed				
6	Reconstruct facilities needed to accommodate the new modules and cells				
7	Install services and communication systems				
8	Prepare any necessary foundations for equipment				
9	Install heating, lighting, air, fume extraction and other services including any specialist environmental equipment				
10	Prepare floor, covering with an appropriate surface				
11	Reinstate and environmentally clean areas of factory no longer required for production				
12	Paint cell boundaries to provide clear identity				
13	Select range of representative components for manufacturing evaluation on new production and test equipment				
14	Establish machine acceptance criteria and conduct trials at suppliers, prior to shipment and installation on site				
15	Prepare machine operating instructions and determine tooling requirements for range of products planned for the machine				
16	Install new and refurbished equipment in the modules and cells				
17	Install ancillary equipment into modules and cells				
18	Determine the management structure and select the module leader/support team leaders				
19	Recruit team members and assess the training needs for groups and individuals				
20	Initiate training in teamwork, cellular organizations, continuous improvement groups, quality procedures and technical skills required to operate different processes				
21	Prepare module and cell quality operating procedures				
22	Release people not required in new structure, making any necessary arrangements for them to train other people in specialist skills				

(Contd)

Table 5.4 *(Contd)*

No	*Work packages Phase 2 – installation*	*Person responsible*	*Start date*	*Finish date*	*Budget*
23	Establish support and assistance for people seeking new employment				
24	Prepare and agree acceptance checklists for handing facilities over to supply-chain operations				
25	Confirm facilities and equipment meets specified performance criteria and the contractual requirements				
26	Conduct process FMEA and confirm possible risks and corrective actions for installation phase				
27	Verify environmental, health and safety requirements for buildings, equipment and facilities fully met by all contractors and suppliers, including documentation and certificates				
28	Agree hand-over dates for modules and phased launch dates for cell teams				

The work packages for extensive building modifications are usually contracted to specialist suppliers who take full responsibility for building work and installation of site facilities. This has the advantage of providing additional resources, reducing the overall time plant and equipment are unavailable for production. However, it is important to compress this phase to the absolute minimum due to the hazard caused to customer deliveries and service while production facilities are out of commission.

Table 5.5 Work packages for phase 3 – implementation

No.	*Work packages Phase 3 – implementation*	*Person responsible*	*Start date*	*Finish date*	*Budget*
1	Officially launch new facilities, handing over responsibility to supply-chain team with ongoing support from project team, equipment suppliers and contractors				
2	Commission new and refurbished equipment using production tooling and documented production processes				
3	Prepare training matrix for formal and on-the-job training for the new processes, identifying suitable instructors				
4	Cross-train people on equipment, formally testing acquired skills before being allowed to assess quality of own work				

(Contd)

Table 5.5 *(Contd.)*

No.	*Work packages* *Phase 3 – implementation*	*Person responsible*	*Start date*	*Finish date*	*Budget*
5	Instigate improvement groups with supporting facilities needed to implement ideas and document changes to processes or working practices				
6	Train team members in quality procedures and problem-solving techniques				
7	Verify capability of key processes and associated measuring systems using production methods, recording any deficiencies				
8	Establish tool management system including methods of servicing and storing tools				
9	Run prove-out trials for tooling, numerical programmes, work instructions and measuring systems for all components, verifying conformance to drawing				
10	Instigate first-line and preventive maintenance procedures, including a machine breakdown log on key machines				
11	Document machine changeover routines, assembling necessary tools and formal practice sessions to improve performance				
12	Evaluate test equipment, checking calibration against formal standards				
13	Establish techniques for changing test equipment to evaluate different parts without impacting production capacity				
14	Agree module and cell performance measures, establishing attainable targets for quality, schedule adherence, stock and productivity				
15	Agree an identity for module and acceptable standards of personal appearance				
16	Establish housekeeping rules and expected levels of cleanliness in modules				
17	Establish the methods for planning and controlling materials in modules and cells				
18	Determine the size of Kanban and containers				
19	Train teams in rules for operating Kanban systems, use of gateways and how to read Kanban cards				
20	Install system for cleaning containers used to protect parts from damage				
21	Establish production targets, increasing production rates to meet manufacturing system design criteria				
22	Determine critical factors impacting cell performance, key quality parameters and methods for implementing corrective actions				

(Contd)

Table 5.5 *(Contd.)*

No.	*Work packages Phase 3 – implementation*	*Person responsible*	*Start date*	*Finish date*	*Budget*
23	Establish management cell audit reviews and instigate corrective actions on features adversely impacting team performance				
24	Prepare a report on the manufacturing system design process, including lessons learnt				
25	Provide manufacturing support to identify and resolve problems in factory or at suppliers, confirming certified products are delivered to the customers				
26	Disband project and installation team, handing over to factory support groups				

Once the modules and cells have been installed and commissioned, the most significant task is to train people in new ways of working, developing the necessary skills to undertake numerous tasks. This invariably takes longer than anticipated, but once people have acquired a broad skill base and become accustomed to working in teams, future design projects are considerably easier to implement. The realisation of improvements in operational performance is sometimes thwarted by key suppliers, so full consideration must be given to improving their effectiveness, bringing them up to similar quality standards and the on-time delivery performance needed in order to commercially exploit the new supply-chain and manufacturing systems.

Investment sanctions

Once the manufacturing system concept design is complete, the project now must gain senior managers' approval to sanction the implementation phases and provide the funds needed to support the proposed capital expenditure and revenue expenses required to introduce the changes. The management authority considered necessary to approve the proposals will depend upon the amount of risk being taken by the company and expenditure required for executing the planned changes.

Project classification

Companies should consider introducing a formal project classification system that categorizes projects into (say) one of four bands based upon the estimated financial investment and level of commercial risk to the business/company identified in the business plan (Table 5.6). This project

Table 5.6 Classification of a project

Company classification – strategic importance/risk to the company	*Business classification – value/risk to business*		
	High	*Medium*	*Low*
High	1	2	3
Medium	2	3	4

categorization is used throughout to establish mandated authorization levels for the project to proceed through various phases.

This classification determines the visibility that must be provided to senior managers and the subsequent control exerted by them. It is also used to determine the level of authority needed to agree the implementation programme, sanction the investments and expenditure limits for various managers (Table 5.7).

Table 5.7 Possible authority levels

Authority level	*Class 1*	*Class 2*	*Class 3*	*Class 4*
Managing director	Mandated	Recommended		
General manager	Mandated	Mandated	Recommended	
Project director	Mandated	Mandated	Mandated	Mandated

The finance director should support this project categorization through a formal investment sanctioning process, allowing senior managers and directors who understand the overall requirements to formally co-sign, authorizing expenditure on equipment and reconstruction of facilities. Information to be collated and presented in order to gain board approval includes:

Summary of the proposal and expected financial returns

- Name of business requesting funds.
- Outline of proposal referring to business initiatives and objectives presented in business plan:
  - reasons why expenditure is required;
  - which business processes will be improved; and
  - which customers and product lines will benefit.
- Costs involved to implement the proposal:
  - capital equipment including tooling and auxiliary items;
  - refurbishment of existing plant and equipment;

  - preparation of facilities and installation of services;
  - associated project and implementation revenue costs;
  - cost of training;
  - redundancy charges; and
  - closure and reinstatement of existing facilities.

- Estimated savings over the life cycle of various products:
  - manufacturing cost reductions on specific product lines;
  - impact on product line margins and profitability;
  - reduction in quality costs;
  - reduction in manufacturing overheads;
  - wages and salaries of people released from company;
  - reduction in space required for production; and
  - closure of facilities due to consolidation and new working practices.

- Cash flow, capital and revenue needed to fund the implementation phase:
  - dates when cash will be committed for expenditure;
  - time scales for expenditure and expected progress payments;
  - cash flow statement for capital items with cash requirement dates;
  - cash flow statement for revenue expenses;
  - grants available to fund capital equipment, reconstruction or training;
  - any prior company provisions for reorganization;
  - rate of return on investment; and
  - discounted cash flow statement for proposed expenditure.

Proposition for the investment

- Introduction, including current business position:
  - brief explanation of how the business evolved, perceived reasoning for present location and how current market position was established;
  - past achievements demonstrating the ability to change and grow market share through exploiting new business opportunities.

- Market background and business potential:
  - critique of market trends and changes that create a need to invest;
  - evaluation of the customer base and their position in the market;
  - size of the potential global market;
  - analysis of the major competitors, including an assessment of their strategy;
  - key buying factors for the range of products subject to supply-chain and manufacturing design;

  - consequences of not investing in the new manufacturing facilities;
  - alternative approaches that have been evaluated and found less attractive.
- Business proposals describing how they relate to the identified business strategy:
  - outline statements included in business plan supporting the business case;
  - reiterate the strategic direction identified for the business;
  - describe how proposals support strategic direction agreed for the business;
  - valid justifications why investment is required and cannot be deferred;
  - statement on how the strategy is enhanced by proposed investment;
  - assessment of strategic and tactical options following successful implementation.
- Quantified performance targets and enhanced financial commitments resulting from the manufacturing system design:
  - statement on savings expected from investment:
    - ◊ wages, salaries, cost of quality, management overheads, working capital, fixed overheads, direct materials and consumable items;
  - target costs, savings and potential gross margins from range of product subject to supply-chain and manufacturing system design;
  - impact upon customer satisfaction and ratings;
  - improvements in non-financial measures of performance that are important to customers and the business:
    - ◊ reduced lead times, enhanced delivery performance, fewer quality problems and a more flexible, responsive supplier.
- Personnel issues to be addressed:
  - job definitions and skills requirements for proposed cellular structures;
  - assessment of skills gap and training needed to introduce cellular manufacturing;
  - methods of providing team training, with time scales and costs;
  - numbers of people required in proposed structure compared to existing organization;
  - analysis of specific skills shortages, stating ways of resolving them, retraining, recruitment, contract employees and such;

  - selection criteria for module and support team members;
  - number of people surplus to requirements, either:
    - ◊ transferred to other jobs on site;
    - ◊ offered alternative employment within the company; or
    - ◊ made redundant and released;
  - management organizational and reporting structures.
- Detailed financial statement produced showing the commercial appraisal and investment justification for the product families included in the supply-chain and manufacturing system design. Information should span an appropriate time scale and be linked to a consolidated overall business report with direct comparisons made to the business plan, highlighting major variances.

Sales

- Sales of original equipment.
- Sales of spare parts and repairs.

  ⇒ total sales

Profit

- Profit on original equipment.
- Profit on spare parts and repairs.

  ⇒ factory margin

Net

- Product introduction expenditure.
- Business development costs.
- Selling and distribution costs.
- Other charges.

  ⇒ trading profit

- Major reconstruction including cost of undertaking manufacturing design, refurbishment, installation costs and training of team members.
- Redundancy costs and relocation packages.
- Closure and reinstatement of facilities.
- Other exceptional items.

  ⇒ profit before interest and tax

- Interest.

  ⇒ profit before tax

Cash flow

- Operating profit before interest and tax.
- Plus depreciation charged against capitalised plant and equipment.
- Capital expenditure needed to implement the change programme.
- Movement in stock and work in progress.
- Movement in operating debtors and creditors.
- Other factors impacting the cash flow.

$\Rightarrow$ operating cash flow

Balance sheet

- Gross stocks.
- Debtors.
- Creditors.
- Progress payments on work carried out.
- Provisions or investments.
- Total capital employed.

$\Rightarrow$ fixed assets

Business ratios

- Stock to sales per cent.
- Stock turns on cost of sales.
- Return on capital employed.
- Return on sales.
- Capital turnover ratio.

Employees

- Full-time staff.
- Full-time hourly paid.
- Part-time employees.
- Contract employees.

$\Rightarrow$ total employees

Employee ratios

- Payroll costs.
- Sales per employee.
- Payroll to sales ratio.
- Added value per employee.
- Added value to payroll costs.

Gross margin analysis

- Original equipment.
- Spares and repairs.

Impact upon non-payroll overheads and costs

- Depreciation.
- Rent, rates and local taxes.
- Gas, electricity, water, fuel oil.
- Consumable materials.
- Maintenance and repair of equipment.
- Insurance premiums.
- Operating leases on plant and equipment.
- Other adjustments.

  ⇒ Total cost of facilities and equipment

Information should also be provided showing

- Rate of return on investment for the proposed expenditure.
- Breakeven point, including maximum outlay and rates of recovery.
- Discounted cash flow statement.

Financial assessment stating and quantifying

- Viability of investment.
- Impact on the business performance and competitive position.
- Impact on borrowings or investments.
- Source of funds.
- Cost of obtaining and securing funding.
- Consequences of not investing.
- Risk assessment and further opportunities:
  - risk analysis associated with proposed changes;
  - actions taken to minimize impact of risks;
  - contingency plans for overcoming potential problems;
  - tactics to protect customer deliveries;
  - opportunities that could arise following a successful implementation:
    - ◊ expand market share due to a lower cost structure;
    - ◊ attack weaker competitors putting them under pressure;
    - ◊ build international strategic alliances;
    - ◊ develop major new accounts.

- Impact on shareholder value and earnings per share:
  - predict movement in shareholder value;
  - impact on stakeholders and possible reactions;
  - key elements of proposed changes promoting shareholder value;
  - propose method for tracking impact upon shareholder value.
- Top level project plan:
  - overall time-phased implementation plan showing key milestones;
  - summary of resources needed to implement the proposals;
  - statement of skills needed to implement changes in planned time scale, also availability of suitably qualified people;
  - possible sources of additional expertise to assist implementation and protect customer deliveries;
  - cost summary:
    - ◊ non-recurring expenditure on people to support implementation;
    - ◊ capital plant and equipment;
    - ◊ revenue expenditure;
    - ◊ implementation and training costs:
  - approval statements from the project manager and supply-chain manager to concur with the proposal, committing to deliver the improved operational performance.
- Conclusions and recommendations:
  - A concise statement should be presented that encapsulates project objectives with a strong recommendation from the local management team that this change programme forms an integral part of the business strategy, and is fundamental to long-term profitability.

Investment plan

- Aims and deliverables from the investment:
  - reasons for designing supply chain;
  - shortcomings of current manufacturing methods;
  - cost competitiveness of present manufacturing methods;
  - manufacturing approach taken by major competitors;
  - summary of the manufacturing strategy outlining concepts to be adopted;
  - aims of new supply-chain and manufacturing systems;
  - quantified deliverables predicted for new manufacturing system:
    - ◊ reduction in lead time;
    - ◊ savings on quality costs;

  - ◊ consistent on-time delivery;
  - ◊ reduced manufacturing costs;
  - ◊ increased product line profitability;
  - ◊ impact upon customer's key buying factors;
  - ◊ increased capacity;
  - ◊ greater flexibility in responding to customer requests.

- Approach taken identifying and selecting appropriate methods and technology:
  - brief description of possible solutions:
    - ◊ full automation;
    - ◊ manual intensive;
    - ◊ hybrid based upon business requirements;
  - assessment of the strengths and weaknesses of alternative approaches;
  - recommended solution with a critical assessment of reasons for making the selection;
  - policy to be adopted and applied throughout implementation.
- Description of areas subject to changes:
  - factory locations involved in design scheme;
  - manufacturing processes identified as core, retained in-house;
  - proposals for organizing supply-chain processes into modules around product families or core components;
  - benefits expected from teamworking and self-managed work groups;
  - statement on current state of equipment and level of process capability;
  - assessment of bottleneck processes;
  - proposals for protecting customer deliveries throughout the transition;
  - involvement of key suppliers and approach taken to supplier integration;
  - support services required to make supply-chain processes effective.
- Items of equipment to be purchased:
  - description of key new items needed;
  - critical features recommended to increase flexibility;
  - standards applied to maintain common features:
    - ◊ tool holders;
    - ◊ tooling suppliers;
    - ◊ numerical controllers;

        ◊ consumable and replacement parts;
        ◊ same supplier for groups of machines/test equipment;
        ◊ environmental, health and safety requirements;
        ◊ electrical regulations;
        ◊ international safety approvals and standard marks (CE);
        ◊ guards and safety shutdown devices;

    - items of plant and equipment needing refurbishment to bring process capability to an acceptable standard;
    - status of measuring systems and level of capability required for controlling critical parameters;
    - additional items of capital expenditure needed to make processes effective.

- Equipment suppliers:
    - matrix of critical features offered by prospective machine/equipment suppliers;
    - minimum equipment specifications needed to perform key operations;
    - indicative prices, terms and conditions for supplying plant/equipment;
    - shortlist of suitable alternative plant and equipment suppliers.

- Capital funding required for items of equipment and facilities:
    - price of basic equipment;
    - additional items needed to bring machines to required specification;
    - tool holders, tooling, jigs and fixtures needed for holding work;
    - automatic loading/unloading devices;
    - measuring systems for verifying and maintaining process capability;
    - module/cell information and control system, including integration with the site computer system for downloading and feeding back information;
    - environmental controls;
    - waste and effluent treatment;
    - foundation work;
    - transportation of equipment to site;
    - installation and commissioning.

- Revenue expenditure to support installation:
    - clearing and preparing the site;
    - coating floors to protect surface, allowing them to be easily cleaned;
    - installation of adequate lighting;

- installation of ventilation/extraction equipment;
- adapting machine instructions and post-processing data for numerical programmes needed for the new equipment;
- prove-out batches on components, using new equipment and tooling;
- storage facilities for tools, work instructions and quality records;
- replacement measuring equipment which is process capable;
- chemicals, oils and consumable items needed to operate equipment;
- safety certificates and third party inspections to approve installation;
- reinstatement of facilities being vacated;
- environmental cleaning of the site, if contaminated.

- Cost of training and identifying suitable providers:
  - assessment of the training costs for:
    - ◊ operating new equipment, including first-line maintenance;
    - ◊ programming equipment and process verification;
    - ◊ working in teams and self-directed work;
    - ◊ production control and Kanban systems;
    - ◊ problem-solving techniques;
    - ◊ continuous improvement groups;
    - ◊ ethical conduct;
    - ◊ quality systems and procedures;
    - ◊ process capability studies and statistical process control;
  - providers of suitable training:
    - ◊ skilled team member providing on-the-job training;
    - ◊ internal training support group;
    - ◊ local colleges;
    - ◊ universities;
    - ◊ research establishments;
    - ◊ specialist training agencies.

- Project schedule to meet key milestones and protect customer deliveries:
  - date for authorization of investment sanction;
  - dates for capital sanctions to achieve committed delivery on equipment;
  - lead times on key items of equipment;
  - dates for scaling down and halting production in preparation for installation of new plant and equipment;

  - time scales for installation and commissioning plant and equipment;
  - module and cell launch dates;
  - dates for identifying and selecting team members;
  - dates for initiating training plans.

- Options on financing arrangements, terms and conditions:
  - outright purchase;
  - operating lease;
  - rent;
  - purchase and leaseback;
  - purchasing terms and cash profiles;
  - warranty periods and rate of response to rectify faults;
  - expected service life and performance guarantees;
  - maintenance agreements;
  - service contracts, items covered and service provided;
  - modifications and upgrades to equipment;
  - acceptance trials;
  - freight costs;
  - on-site installation and commissioning;
  - on-site training on use of equipment;
  - payments for additional training;
  - project management of systems integration and interfaces with other equipment.

- Summary statement on proposed methods for funding various project costs; linked to dates when expenditure will be committed and cash required.

Equipment selection rationale for essential equipment items

- Current methods and procedures for producing or testing components:
  - summary of technical policy for manufacturing;
  - process definition identifying critical product features that must be controlled;
  - shortcomings and deficiencies embodied within current process;
  - customer expectations to remain competitive and sustain future business;
  - alternative processes requiring capital investment in new technology.

- Specific technical options for key items of plant and equipment:
  - review of possible technical solutions;

- risks associated with particular technologies;
- technical and process reliability assessment of alternative manufacturing or processing methods;
- strengths and weaknesses, including calculated production costs;
- technical features regarded as important to ensure full process capability and reliability;
- matrix of essential and desirable technical features offered by different equipment suppliers.

- Recommended preferred option:
  - reasons for selecting a particular process;
  - list of attributes confirming preferred solution meets all the specific process requirements;
  - identify equipment suppliers capable of providing technology;
  - review of equipment suppliers meeting these requirements, assessing technical competence, financial viability, location of field service support and previous experience of vendor;
  - list of preferred suppliers with reasons for making selection.
- Outline machine and equipment specification:
  - establish the operating envelope for the equipment;
  - provide technical specification, defining key performance parameters and technical issues needed for detailed price negotiation;
  - statement of functional requirements needed in production environment;
  - specify expected life of equipment, status of technology and methods of upgrading or retrofitting additional features.
- Operating tolerances and process capability:
  - define critical dimensions and features used to determine process capability and subsequently control production quality;
  - assess repeatability, accuracy and capability of new process compared to former one;
  - specify features to be used for acceptance trials prior to accepting handover of equipment;
  - establish performance guarantees to be negotiated in the contract.
- Total costs for installing specific machines or processes:
  - price of basic equipment;
  - additional items needed to bring the machine to specification requirements;
  - tool holders, tooling, jigs and fixtures;

- automatic loading and unloading devices;
- measuring systems;
- integration into module and site computer system;
- environmental controls;
- waste and effluent treatment;
- clearing and preparing the site;
- foundation work;
- coating floors to protect surface;
- transportation to site;
- installation and commissioning;
- installation of lighting and ventilation/extraction;
- adapting and post-processing data for numerical programmes;
- prove-out for components using new equipment and tooling;
- storage facilities for tools, work instructions and quality records;
- replacing defective standard measuring equipment;
- chemicals, oils and consumable items needed to operate the equipment;
- safety certificates and third party inspections to approve installation.

• Delivery lead times required by suppliers:

- statement on time scales needed by equipment suppliers:
  - ◊ agreeing final specification;
  - ◊ negotiating contract;
  - ◊ designing equipment to meet specification requirements;
  - ◊ procuring parts and building equipment in their factory;
  - ◊ testing and confirming that performance meets the acceptance criteria;
  - ◊ shipping and delivery to site;
  - ◊ installation;
  - ◊ commissioning trials and tooling prove-out on site;
  - ◊ hand over to production.
- key dates needed to meet milestones stated in the implementation project plan;
- contingency arrangements, enabling unplanned changes to projected time scales to be accommodated.

This information collated for the investment sanction may appear overdetailed, but it gives the local management team and company directors a comprehensive overview of the business, together with financial implications for the project. This is essential if the project involves taking significant risks in transforming the fundamental business operation or making large capital investments in new equipment and facilities. *Remember also, if particular items of expenditure are not identified they will still require funding, making the local management team apply for additional finance and reveal limitations in their ability to deliver financial commitments.* The different aspects of this investment sanction are usually reviewed and scrutinized by appropriate group directors responsible for approving the project, prior to the salient points being summarized in a formal paper presenting the business case for making the investment. Well-prepared presentations should aim to promote the benefits of the proposed changes, imparting confidence that the investment is technically sound and can be financially justified within the business's investment criteria. The time and effort required to gain formal approval should never be underestimated, because companies have many demands and potential investment opportunities to consider. It is the responsibility of the local and group management teams to demonstrate that this particular manufacturing solution offers the greatest short-, medium- and long-term benefits to the company and its stakeholders.

Formal approval for investment and implementation

The process for obtaining formal approval to spend the funds and implement the proposals identified in the investment sanction is determined by the individual requirements of the managing director, organization type, company size, the chairman and major outside investors. In my experience, larger companies usually require the relevant information collected for the investment sanction to be summarized as a formal board paper for consideration by the company's main board of directors. The investment sanction will have been evaluated prior to submission by the appropriate group directors responsible for investment authorization, in this instance the supply-chain director, finance director and – where necessary due to the level of business risk – the company managing director. The local management team applying for the funds is required to attend the next company board meeting, making a case for the directors to sanction the investment. The team should make a considered and well-crafted presentation, followed by discussion, allowing directors to raise relevant issues and evaluate the overall business case. Once the ramifications have been fully considered and agreement reached by the main board to support the proposed investment, the board paper and investment sanction are signed by the managing director, providing authority for the project to proceed through to implementation. This main board review is the most significant milestone as it financially commits the company to making the investment, releasing funds to implement the proposed changes.

Capital sanction approval

Once the investment sanction has received board approval, the next stage is to prepare individual capital sanctions for significant items of capital equipment. This formal documentation is usually signed by local and group management teams, giving people authority to raise purchase orders for new equipment and implement site preparation work. The capital sanction documentation should be user-friendly but structured to prompt answers to all relevant technical and commercial questions that must be addressed prior to signing a purchase order with the supplier. Information requested should be comprehensive but tailored to satisfy the particular business needs. Consideration should be given to the following.

Front cover of the investment sanction

- Brief summary report of the salient proposal features.
- Significant risks associated with project.
- Key financial figures showing investment required and identified savings.
- Cash flow and rate of return.
- List of managers requesting funds with signatures and dates.
- Statements from managers responsible for implementing the recommendations confirming they champion the proposals and require the funds to meet business plan commitments delivering enhanced operational performance.

Technical specification, requirements and associated costs

- Detailed machine, plant or equipment specification.
- Special options and auxiliary equipment.
- Tooling and fixtures.
- Material handling facilities and pallets.
- Removal of waste materials and solvents.
- Storage for tooling and pallets.
- Programming of parts needed for commissioning.
- Re-engineering and tooling for existing components.
- Foundations and load-bearing requirements.
- Environmental requirements and levels of cleanliness.
- Power supply, cooling system and other services.

- Switch gear and electrical connections.
- Control systems, computers and communications.
- Software, translators and post-processors.
- Interfaces to existing equipment.
- Agreed performance criteria for acceptance, including capability guarantees.
- Integration of equipment into existing systems.
- Warranty agreements and expected service life.
- Service contracts and time intervals between servicing.
- Service items to be purchased, including chemical substances.
- Penalties for late delivery or failing to meet performance criteria.
- Negotiated cost of total package.
- Agreed delivery date.
- List of equipment being replaced.
- Current book value of equipment and tooling being replaced.
- Allowances and trade-in value of existing equipment.
- Costs associated with removal and disposal of old equipment.
- Payment options available from supplier and leasing alternatives.
- Preferred terms and conditions of payment.

Additional items that should be addressed

- Control and disposal of waste materials and chemicals.
- Health and safety requirements for operating process.
- Noise levels and acoustic screening requirements.
- Conformance to international standards for machine safety.
- Responsibility for installation of equipment.
- Removal and future use or disposal of existing equipment.
- Area of factory space released or required.
- Responsibility for machine commissioning and associated costs.
- Running times and proposed utilization of equipment.
- Agreed changeover times.
- Measuring systems to be used in production.
- Routine calibration of equipment and measuring systems.
- Process capability of measuring system and critical parameters of components.
- Routine maintenance requirements.
- Proposed preventive maintenance schedule and responsibilities.
- Packaging and freight costs.
- Removal of packaging and materials used for commissioning.
- Preparation and refurbishment of facilities.

Revenue expenditure to support the investment

- Training needed to operate the equipment.
- Material required for training and trials.
- Relocation of existing plant and equipment.
- Closure and reinstatement of buildings.
- Disposal of existing liabilities.
- Environmental and ground surveys to confirm site is not contaminated.
- Preparation of operating procedures and quality systems.
- Training in quality procedures.
- Obtaining third party quality approval for products and processes.
- Licences to operate equipment.
- Regular upgrades to keep the equipment current with latest standard.
- Cleaning procedures and disposal of effluents.
- Service contracts and agreements.

Expenditure request

- Summary of the performance specification.
- Special options and auxiliary items of equipment.
- Related factors that must be dealt with as part of the investment:
  - formal quotation giving the price of capital equipment items;
  - cost of supplementary items;
  - revenue expenditure required to support the investment;
  - cost of training:
    - ◊ *Total expenditure request;*
  - cash flow statement.

Savings and returns

- Saving to be secured by the investment.
- Payback period in months.
- Rate of return on investment.
- Date of order commitment.
- Date of payment.
- Discounted cash flow statement for the investment.
- Sale price of existing equipment.

The information requested needs to be relevant to the business and appropriate to the type of equipment being purchased. However, introducing a discipline to consider the broader operational requirements and details of associated expenditure ensures these factors are fully considered prior to obtaining an investment sanction. *It is dishonest and misleading to request funds for base capital equipment alone,* as these costs are considerably lower than overall implementation costs. It is important all aspects of expenditure are taken into account in the capital sanction application, because these additional funds are vital for ensuring processes will be fully capable, operating to specification. It also saves the awkward and possibly 'career-limiting' situation of having to raise supplementary sanctions for additional expenditure to complete the job.

The capital sanction should be raised and completed by the team of managers who require the investment, supported by the site general manager. This is subsequently authorized by the finance director, programme director, purchasing manager and managing director. However, the process must be organized to allow sanction requests based upon an agreed business plan and authorized investment sanction to be approved within a few days. Once the investment decision to proceed has been made it must not be subject to further bureaucratic delays that could jeopardize achieving key milestones.

Summary

The business planning process, if performed correctly, should have ensured that the managing director and his team are fully aware of the business objectives and understand the financial challenges facing a particular business. The capital investment requirements and commercial risks therefore should have been considered and agreed in principle by the managing director at the outset of the project. Companies should consider introducing formal project classification systems that categorize projects based upon the estimated financial investment and level of commercial risk identified in the business plan. This is then used to establish the level of authorization needed for the project to proceed.

The process for obtaining authority to implement the project team's proposals is shown in Chart 5.4.

Preparing an investment sanction and obtaining formal local and main board approval are lengthy procedures requiring considerable information to be collated. It appears to add little value to the proposal, but it is worth determining which items of information are irrelevant when directors are faced with making important investment decisions that involve considerable financial risk. In my experience, presenting the technical and business case to experienced senior managers means also that ideas and assumptions are rigorously challenged, leading to more realistic and workable solutions. It must always be remembered that companies have

Chart 5.4 Authorization process

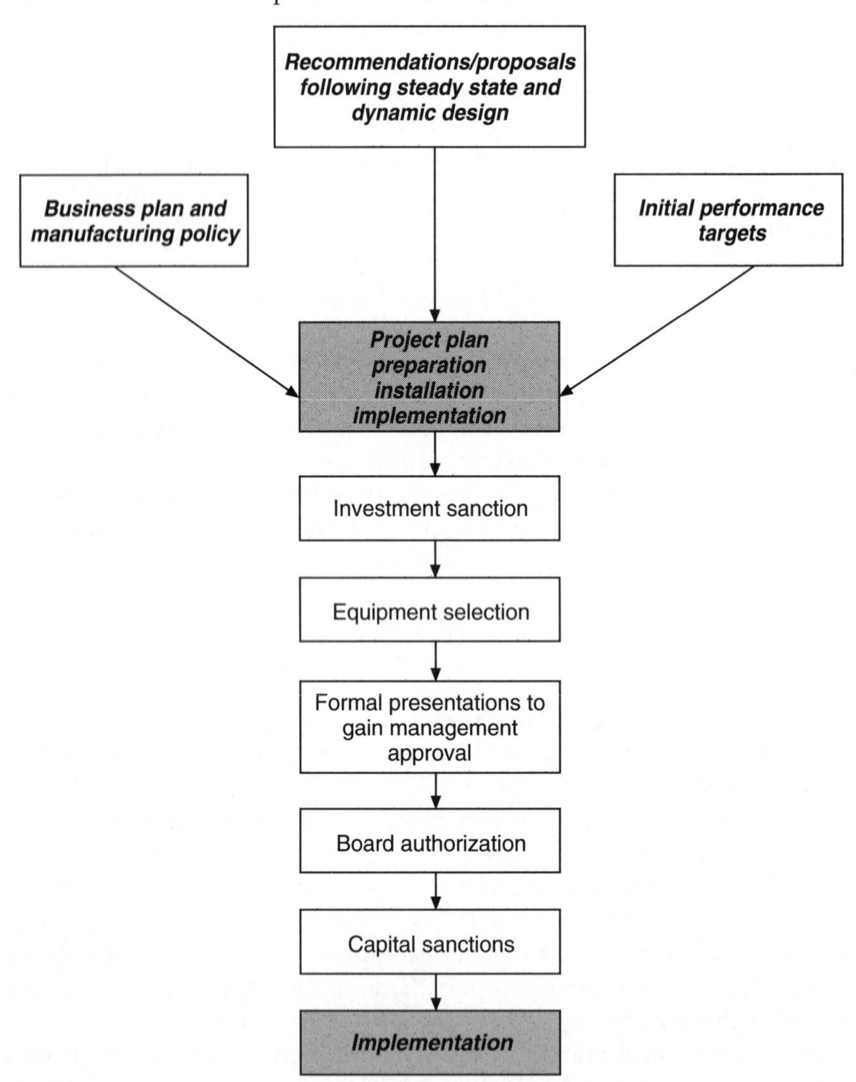

finite funds that must be deployed in the best interests of both shareholders and stakeholders. Company directors must make difficult decisions on where to invest valuable management and financial resources, but it is incumbent upon them to ensure their businesses remain viable both now and in the future.

> *This crucial project milestone is the last opportunity to make changes without incurring major time and cost penalties. So, once authorized, the plan is to implement the proposals as rapidly as possible!*

Chapter 6

Commissioning and continuous improvement

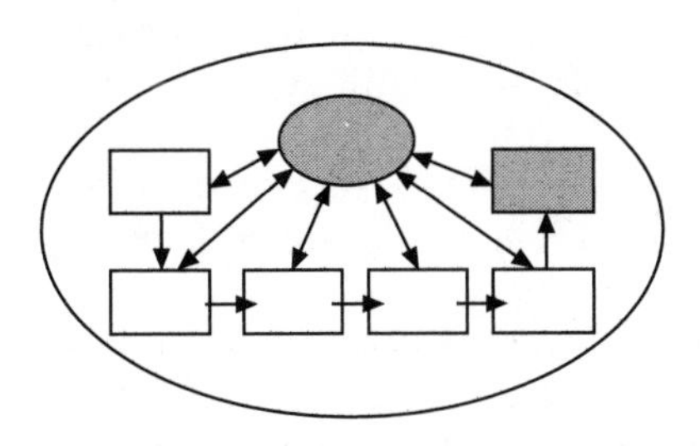

Module implementation/case study

Once the authority to proceed has been obtained and all the detailed preparation work completed, the final phase is to implement the factory reorganization in as short a time as possible, minimizing disruption to customer deliveries. Whenever possible, significant reconstruction should be planned to take place when the factory is closed for holidays or in periods when workloads are traditionally lighter. Specialist contractors experienced in moving facilities and installing services should be used, providing the additional benefit of extra resources needed to complete the factory installation quickly. The project management and sequenced planning of events throughout the reorganization has a significant impact upon the overall time taken to complete the moves. The aim should be to systematically move all the equipment associated with a particular cell, progressively handing it back to production, focusing on the bottleneck processes that cannot be replicated on site or at subcontractors. Senior managers should consistently review progress, forcing the pace and taking immediate action on unplanned events that may cause delay or disruption.

It is impossible for the project team to have designed an optimum manufacturing process, due to the sheer complexity of the task and the number of variables that were taken into account. The initial design now requires the implementation team and module/cell team members to work together to refine the process, introducing a continual improvement process aimed at manufacturing consistently good quality products, on time and eliminating waste from all aspects of the process. Over the past few years many companies have reaped the benefits from using continuous improvement teams; this approach must be championed and rigorously promoted by senior managers to transform the supply-chain and manufacturing system design into a world class process. It should also be understood that if a manufacturing system has simply *evolved*, without a *systematically designed* production process with associated support services, the chance of developing an optimum manufacturing process by means of continual improvement teams is extremely low.

> *To be world class, the manufacturing system must be systematically designed and further developed using continuous improvement teams.*

In my opinion the design of the supply-chain and manufacturing systems is fundamental to creating effective factories. The Japanese have demonstrated these principles in factories worldwide, but the West continues to resist putting the necessary level of resources into designing and implementing such systems.

A case study

The initiation of any new manufacturing system is always a difficult period and requires leadership and persistence to re-establish the smooth operation of tasks fundamental to running the business. I will next describe an actual situation of which I have personal experience where a manufacturing system was redesigned to change a business's cost base from a traditional military contractor into a competitive civil aerospace business, capable of meeting contracted price commitments. Despite an extensive manufacturing systems design programme and capital investment in the best possible equipment, the process simply failed to operate. The result was several new aircraft were lined up on the tarmac awaiting parts, with irate intermediate customers and airlines seeking to invoke expensive penalty clauses for late aircraft deliveries. I assigned myself the role of acting operations manager, becoming leader of the factory implementation team responsible for 'making it happen'. In this situation the only way forward was to go back to fundamentals!

The first step was to assemble the management team, holding an open workshop to gain a common understanding on the seriousness of particular problems and work together to establish an action plan on how to rapidly recover customer delivery schedules.

The agreed plan had the following stages, many of which happened in parallel:

- Assign the most appropriate manager to work full time with module leaders, identifying issues that were restricting production and causing customer problems, including those that could not be resolved by the module/cell team. This made product line managers directly responsible for the operational performance of modules manufacturing their customer's products; the chief metallurgist was assigned to run heat treatment and processing cells, and the chief production engineer took direct control of all maintenance activities.
- A presentation was made to the workforce, telling them in simple terms:

  - the 'hurt' we were inflicting on our major customer;
  - the size of the problem, based on the number of units and monetary value to us and the customer in lost sales;
  - actions our customer might take to resolve our poor performance;
  - management's assessment of the situation if we lost this business;
  - actions we, as managers, intended to take to recover the situation;
  - what was expected of everyone to demonstrate how we valued our customer and would strive to remain as a preferred supplier on future contracts.

  This was followed by an open question and answer session to clarify specific issues.
- A study was made of the material flow through the supply-chain process, quantifying the level of stock at each holding area. Bottlenecks were found at:

  - incoming materials stores;
  - goods inwards inspection;
  - heat treatment;
  - heat treatment testing, material approval and release;
  - kitting stores prior to assembly for products still assembled in the general assembly area.

  The quality manager was assigned the full-time task of removing these bottlenecks and reporting daily progress.
- The production and commercial managers were 'confined' to an office and instructed to establish a realistic production schedule that

recovered the customer deliveries in a sequence and time scale that our customer could reluctantly accept, with the following provisions:

- every item manufactured must go to a customer;
- any items on the schedule must have all bought-out components available in time for manufacture;
- the production schedule would not exceed the accepted plant capacity by more than 20 per cent, being resourced by additional overtime and Sunday working;
- once agreed, manufacturing would be committed to meeting the production schedule; any changes would need a formal written request from the customer.

- The recovery programme was managed using a mandatory daily 12 o'clock meeting scheduled to last for half an hour. (If overrun, people were in danger of missing lunch in the canteen!) The managers responsible for key activities and module managers attended this meeting and collectively we tabled all the items that needed management resolution. Action lists were prepared and circulated within two hours of the meeting giving a description of the task, person responsible and committed completion date. Progress towards accomplishing these tasks was regularly reviewed and support given to ensure the committed dates were achieved.
- The factory performance report was compiled weekly and modified to provide daily information on the work produced within each cell. The figures were adjusted, as requested, making an allowance for traditional in-process activities to be included. I introduced a moving base line that automatically increased whenever cell teams exceeded their previous best performance; I walked round the factory acknowledging people's achievements and saying *thank you!* In fact I spent over 60 per cent of my time walking through the factory, picking up problems first hand from the people affected by them, making commitments to resolve issues whenever possible, or simply explaining why the situation could not be addressed immediately.

One morning during the first week I was stopped by a senior union representative, who asked, 'Why are you walking the factory so early in the morning?' 'To say good morning!' was my immediate response. This was suspiciously accepted; three weeks later I met this same person and asked, 'Is there a problem with me walking the factory in the morning?' and his spontaneous reply was 'Oh no, we miss you, if you *don't* come!'

Using this planned approach took four months to re-establish a stable relationship with the customer; factory output increased by 2 per cent each week, giving a total increase of 33 per cent. This was a significant step towards achieving the 50 per cent improvement goal that had been established for the manufacturing system redesign. It is unlikely these dramatic performance improvements could have been delivered using traditional manufacturing methods; the new system provided the platform for further improvements as cell teams started implementing continuous improvement programmes. Finally I must pay tribute to the factory workforce who throughout the crisis provided full support and worked incredibly long hours with great commitment to maintain the confidence of our customers, proving that people are indeed the most valuable resource available to a company.

Machine and process capability

Installing equipment in the cell can be accomplished quickly; a longer task is to confirm all items of equipment, including test facilities, are process capable and will consistently perform operations within specified tolerances. It is important to refurbish existing equipment replacing any worn parts, and confirming all machines are process capable as part of the acceptance trials, prior to installation in the new cell.

> *Process capability is fundamental to creating effective manufacturing systems because if machines are not process capable they produce higher levels of rejects, disrupting the flow of work so destroying many of the benefits to be gained from modular manufacturing.*

Established methods for demonstrating machine capability are:

- Measure the key characteristics of a number of machined components (ideally a sample greater than 40) without making adjustments, plot the spread of tolerances due to variations in the machine, material or tooling and use standard routines to calculate the process capability (Cp).
- Obtain the manufacturer or specialist machine repairer's inspection reports, giving detailed information on wear in slide-ways, spindles, axis alignment and concentricity of rotating elements.
- Use precontrol charts to verify that 5 components can be made within 50 per cent of the total tolerance band. This technique demonstrates the process is incapable unless all five are within tolerance. (Quick method, developed by Peter Shainin and described in Bhote (1987).)

If equipment is incapable of operating to the required level of process capability, then only four management choices remain:

1 Confirm with the product introduction team the reason for specifying such tight tolerances and if possible reconfigure the part, relaxing those tolerances on the drawing or material specification that allow the part to be made to drawing.
2 Repair or refurbish the machine to re-establish process capability.
3 Buy a new machine or equipment that is capable of operating to the required tolerances.
4 Manufacture the item using a different process, either in-house or using a subcontractor.

It is difficult to assess the overall capability of an assembly process, even if component parts can easily be taken apart and reassembled. Isoplot techniques can be used to isolate particular aspects of variability, using two measurements taken from several components and plotting them against each other to show the degree of repeatability (Peter Shainin technique, Bhote (1987)). However, in most circumstances provided that assembly equipment is calibrated and verified to be in good working order, the quality of component parts is the determining factor in supplying high quality products.

The calibration and process capability of measuring systems also have a significant impact upon the overall process capability and these must be *an order of magnitude* more precise than the tolerances being measured. Deficiencies in the measuring system are significant, because they erode the available processing tolerance and are generally less expensive to correct than actual process deficiencies.

The suitability of measuring systems should be confirmed by:

- Gauge Repeatability and Reproducibility (Gauge R&R) studies to determine the accuracy and repeatability of the measuring systems.
- Isoplots using two measurements taken from several components – measuring system capability is confirmed if the parameters fall within defined boundaries. A quick capability assessment using Isoplots can be obtained by:
  - measuring the same feature in the same position on a batch of 5 components, selected in random order;
  - repeating the measurement on the same components in a different order;
  - recording for each component the first reading against the second one;
  - plotting the first reading against the second one on equal scales;
  - drawing a line at 45 degrees from the intersect point of the two axes;
  - setting the measurement variation ΔM to 1/6 of the tolerance of

the parameter being measured and draw two lines ΔM apart, equidistant from the average line;
- setting the control limits to half the measurement variation ($1/2\ \Delta M$) and draw two more lines equidistant about the 45 degree line.

A measuring system passes the Isoplot test if 5 points lie within the control limits and is suitable for measuring the selected parameter. If the measuring system fails, then further evaluation is required using the full Isoplot method or Gauge R&R. (Isoplot techniques – developed by Peter Shainin.)

The purpose of machine capability studies is to verify process repeatability and consistency; they have limited use for identifying the cause of poor capability or corrective actions needed to improve processes. Therefore, if processes are to be made more robust, action logs are needed for recording faults, experiments designed to identify the root cause of problems and specific actions taken to improve overall process reliability. These routines should be established in conjunction with continuous improvement programmes that are needed to optimize the manufacturing system, following module implementation.

Process capability is at the core of good quality products; no manager can justify allowing module/cell teams to work with equipment that is incapable of producing components consistently to specification.

Job definitions and skills requirement

The selection of people with the correct attitude and skills needed to fulfil manufacturing roles is a very important task. An unfortunate consequence of designing effective manufacturing systems is that they require fewer people than before. For an existing site with an established workforce, unless the site expands its production capacity by increased volume or introducing new products, people will have to be released from the company. A reduction in employee numbers is inevitable, as a significant proportion of the projected cost savings will be associated with employment costs. Methods for selecting people for particular job roles must take into consideration the retention of those individuals possessing appropriate skills, and also must be seen as fair by people losing their jobs. The task of selecting people is equally important when a business needs to expand its local workforce due to introducing new product lines or merging operations.

The role of the module manager is broad, owning resources and being directly responsible for achieving customer quality and delivery commitments. Tasks undertaken include:

- Initiate and implement improvements in customer service, product quality and module operations.
- Introduce change to enhance delivery performance and reduce manufacturing costs.
- Manage resources, allocating appropriate tasks to team members, making allowances for training, continuous improvement groups and first-line maintenance.
- Control the use of equipment, ensuring bottleneck processes are given priority and time is allocated for preventive maintenance.
- As a member of the manufacturing system implementation team, confirm module and cell activities, verifying resources allocated to manufacturing teams and support activities.
- Select and recruit people for modules/cells, ensuring they have correct attributes to work effectively in teams.
- Develop team working skills of individual team members and promote professionalism to enhance overall performance.
- Assess the abilities of team members, identifying the skills gap and assigning people to appropriate tasks.
- Plan the training requirements of team members, providing opportunities necessary to allow people to achieve their full potential.
- Evaluate the performance of individual team members and recommend appropriate remuneration packages based upon skill and acceptance of responsibility.
- Develop interpersonal relationships and create a working environment making people proud to be part of the team.
- Seek information, identify problems and take corrective actions to prevent disruption and unplanned events.
- Monitor performance, providing management information and feedback to module/cell teams on achievements.

The role of module/cell team members must also be expanded. Several companies have acknowledged this by creating new broader job definitions such as 'cell craftsman', 'supply-chain technician' or 'manufacturing associate'. The objective is to establish self-directed workgroups who select their own leader to coordinate activities. Everyone is encouraged to participate fully in developing and enhancing team performance, working within the following framework:

- Contribute to implementing change, improving customer service and product quality.

- Identify and support the most effective job role for people, maximizing their contribution to team performance.
- Recommend ways of improving quality, increasing productivity, utilization of equipment and use of resources.
- Participate in selection and training of team members.
- Contribute to continuous improvement groups, implementing recommendations considered beneficial to individual or team performance.
- Agree and commit to achieving customer schedules on time.
- Seek to continually enhance team performance and train people when required to develop new skills.
- Collectively plan work to be carried out by team.
- Take responsibility for quality of own work and that produced by team.
- Build strong working relationships with other team members and service groups.
- Seek and prepare information needed to report achievements and recommend areas for improvement.
- Agree to work in a flexible manner to support overall team efforts.

The combination of knowledge and skills needed to become a manufacturing technician may include:

- Operating and controlling numerical machine tools and test rigs.
- Installing and proving-out machine programmes for operating equipment.
- Fitting skills to perform additional operations on machined components.
- Assembly and fitting skills to assemble products.
- Knowledge and ability to test products against performance specifications.
- Product knowledge to identify possible causes of failure.
- Continuous improvement techniques and methods for eliminating waste.
- Rapid changeover of tooling.
- Material control systems and the use of Kanbans.
- Materials handling techniques preventing damage of parts in transit.
- How to package and dispatch products ready for use by the customer.
- First-line and preventive maintenance techniques.
- Formal quality systems and documented company procedures.
- Computer literacy and the application of standard IT systems.
- Operator self-inspection methods and associated documentation.
- Tool management systems and control.

- Statistical methods for confirming process capability.
- Standard problem-solving techniques.
- Team working and interpersonal relationships.
- Identifying and understanding customer needs.
- Defining processes and manufacturing systems awareness.
- Health, safety and environmental standards, legal requirements and the need to maintain a safe working environment.

The likelihood of people having an appropriately broad range of knowledge and associated skills is remote. The module manager must adopt a rigorous selection process, ensuring people have sufficient knowledge to adequately perform specific tasks and are able to acquire the range of skills needed to develop into a key module team member. Part of the selection process should include an assessment of immediate training needs, so must be completed before launching the cell team. The module manager must also determine the breadth of education people should undertake as part of their personal development plan. The training commitment to prepare people for working in this broader role should not be underestimated, allowing people sufficient time to acquire working competency. Once training has been given, subsequent changes are considerably easier because team working experience and indigenous knowledge allow new methods to be understood and adopted much more quickly. Some Japanese companies report that it takes over ten years for everyone to become fully competent and complete their training programmes!

Team selection

Module and cell teams are mostly selected from the existing workforce, so it is essential to have a formal selection procedure capable of identifying overall attitudes and skill levels. Correct team selection ultimately determines the success or failure of any future manufacturing system.

> *People make the difference and teams of people working together can resolve problems, engineering processes into world class manufacturing solutions.*

It is no longer sufficient to select people on technical skills alone. As can be seen from the job roles, the requirements are much *broader*, depending upon people being flexible, willing to learn new skills and support the team to achieve its commitments. Apart from the few people recruited for their specialist knowledge, the existing workforce must be trained to adopt these new working practices. Considerable effort must be devoted to selecting people, assessing training requirements and creating balanced team dynamics. One technique used is based upon identifying core competencies. Behavioural competencies that should be evaluated and graded when selecting manufacturing technician team members are:

- *Determination for excellence* – focusing upon self-motivation to achieve results and showing initiative to complete task.

  1. Has little energy or no energy.
     Always needs more time or resources.
     Blames other people.
     Waits for instructions.
  2. Strives to perform the tasks.
     Offers solutions to problems.
  3. Uses own initiative to complete the task.
     Knows how to overcome problems.
     Has high level of energy.
     Has personal credibility.
  4. Self-starter takes initiative.
     Identifies personally with the job.
     Makes all-out efforts, ensuring tasks are completed.
     Accepts responsibility for errors.
     Takes corrective action to rectify mistakes.
  5. Totally committed to achieving excellence.
     Never fully satisfied, thinking of improvements.
     Works to overcome personal shortcomings.

- *Concern for detail* – understanding the detailed requirements, reducing uncertainties and process variability. Planning work routines, ensuring materials, tooling and machines will be available to complete the work on time.

  1. Misses commitments.
     Fails to check documentation/makes mistakes.
     Expects tooling and equipment to be available.
  2. Methodical and thinks about the task.
     Develops established routines for performing tasks.
     Records and logs events that cause problems.
  3. Develops own method for completing tasks.
     Considers alternative techniques for improving effectiveness.
     Establishes priorities and sequences for delivering own work.
     Strives to meet commitments.

4. Fully understands the requirements and shares ideas with team.
   Checks that items are correct before starting a job.
   Identifies problems and takes action to rectify them.

5. Develops and implements continual improvements.
   Continually improves overall process capability.
   Maintains concise/accurate record and log of events.
   Instigates improvements with team members.
   Provides information to people supporting the cell.

- *Team participation and leadership* – ability to be an effective team member, accepting responsibility for motivating people and providing leadership.

1. Will not cooperate with other members of the team.
   Loner, wanting to do things on their own.
   Does not participate in team initiative.
   Works in own style.

2. Seeks to create harmony in team.
   Supports team initiatives.
   Acknowledges the contribution of other team members.

3. Motivates team toward a common objective.
   Understands how people behave.
   Takes people's possible reactions into account.
   Works for benefit of team.

4. Accepts role of leader when required.
   Drives team to succeed.
   Identifies unacceptable performance.
   Provides support and instruction for team members.
   Proud to be a team member.

5. Champions the team and its achievements.
   Shares vision for improving team performance.
   Role model for team.
   Totally committed to delivering enhanced performance.

- *Self-confidence* – belief in own judgements and willing to take responsibility for delivering commitments.

1. Feels other people control events.
   Always asks for confirmation or instruction.
   Does not feel able to take a decision.

2. States opinions clearly.
   Says what is required when asked.

3. Questions non-value added tasks and instructions.
   Has the confidence to act on own initiative.
   Seeks rapid resolution to queries.

4. Takes responsibility for resolving issues.
   Makes decisions that benefit team performance.
   Prepared to back own judgements.

5. Stands up publicly for beliefs.
   Confronts difficult issues constructively.

- *Manages pressure* – ability to maintain efforts under continual pressure to perform tasks and keep emotions under control.

1. Feels unable to cope.
   Experiences constant pressure.
   Anxious about performance.

2. Controlled under pressure.
   Resists acting without thinking.
   Able to cope with provocation.

3. Gives appearance of being in control.
   Has own way of coping with pressure.
   Has outlet for stress.
   Checks negative emotion, talks calmly.

4. Remains optimistic.
   Plans ways of handling pressure.
   Manages expectation of others.
   Clarifies difficult situations.

5. Calms situations.
   Absorbs and relieves pressure from team members.

- *Problem solving* – ability to absorb information applying it to make informed and timely decisions.

1. Slow to understand facts.
   Fails to see related factors.
   Reacts to problems after event.

2. Considers available information.
   Asks appropriate questions.
   Thinks logically.

3. Understands relationship between events.
   Provides workable recommendations.
   Rapidly understands critical elements.
   Knows when to seek other opinions.
4. Able to critically assess complex situations.
   Able to analyse voluminous data.
   Enjoys taking decisions.
5. Consistently makes the correct decision.
   Finds optimum solutions to complex problems.

The minimal requirement for each factor can be established for particular cell teams and candidates rated as part of the interview process, providing additional information for the selection process.

Other important competencies to consider for cell team member selection are technical skills, product knowledge and proficiency in manufacturing techniques. The skills needed to become members of different cell teams have to be established and minimum requirements determined for the various manufacturing activities, taking full account of the module and cell structures being introduced. The level of present ability should be assessed and graded on the following basis:

1. No knowledge of the process.
2. Some awareness, can perform task with close supervision.
3. Can perform the task competently.
4. Experienced and capable of verifying own work.
5. Capable of training others in operational process.

The range of skills selected for evaluation must be tailored to the cell requirements, for example:

Complex machining cell

- Operating:
  - full range of conventional machine tools;
  - numerical control turning centre;
  - numerical control prismatic machining centres;
  - numerical control grinding equipment;
  - gear cutting machines;
  - specialist machines.
- Installing tools and setting up equipment.
- Making modifications to numerical control programmes.

- Creating numerical control programmes.
- Designing and initiating machining trials for new products.
- Selecting and recommending appropriate tooling methods.
- Performing fitting and deburring operations.
- Cleaning equipment to remove debris and contamination.
- Operating inspection systems.
- Developing and verifying inspection routines.
- Understanding quality system and operating procedures.
- Developing and documenting standard working routines.
- Resolving quality problems and instigating corrective actions.
- Completing documentation to verify conformance to drawing.
- Modifying planning routines, improving the flow of materials.
- Recommending manufacturing layouts and sequence of operation.
- Applying statistical process control.
- Verifying capability of measuring systems.
- Undertaking first-line maintenance requirements for equipment.
- Controlling flow of materials and Kanban systems.
- Protecting components from physical and environmental damage.
- Reducing changeover times for equipment.
- Operating the tool management system and controlling gauges.
- Operating computer systems and information technology.
- Applying standard software packages to perform specific tasks.
- Understanding legal, environmental, health and safety requirements.
- Meeting customer requirements and delivering commitments.
- Interpreting the module and cell measures of performance.

Assembly and test area

- Understanding:
  - the build structure and components needed for different products;
  - customer requirements and critical product features;
  - product applications and overall performance requirements;
  - product test specifications;
  - failure modes and corrective actions to prevent possible errors.
- Cleaning and component preparation requirements prior to assembly.
- Sequencing components for assembly process.
- Inspecting components to confirm quality standards.
- Tooling for assembling parts and subassemblies.
- Assembling products, taking account of build specifications and quality requirements.
- Setting up equipment to aid the assembly process.
- Calibrating tools and equipment used in assembly process.

- Selecting individual components needed to meet specific tolerance requirements.
- Fitting components and adjusting fine tolerances to meet specifications.
- Setting up inspection equipment for evaluating products.
- Confirming measuring system calibration.
- Establishing overall capability of the measuring system.
- Performing complete sequence of tests.
- Completing documentation and test certificates.
- Stripping assembled units.
- Diagnosing performance deficiencies and making complex adjustments.
- Recommending modifications to improve design integrity.
- Building fixtures to aid assembly process.
- Developing self-checking methods to prevent possible assembly errors.
- Controlling flow of materials and Kanban systems.
- Planning the build sequence to optimize availability of facilities.
- Developing and verifying inspection routines.
- Understanding quality system and operating procedures.
- Developing and documenting standard working routines.
- Identifying quality problems and instigating corrective actions.
- Completing documentation to verify conformance to specification.
- Modifying planning routines, improving flow of materials.
- Recommending assembly layouts and sequence of operation.
- Applying statistical methods to confirm performance trends.
- Verifying the capability of inspection and measuring systems.
- Undertaking first-line maintenance requirements for assembly and test equipment.
- Reducing the changeover times for equipment.
- Managing assembly fixtures and controlling gauges.
- Operating computer systems and information technology.
- Applying standard software packages to perform specific tasks.
- Understanding legal, environmental, health and safety requirements.
- Meeting customer requirements and delivering commitments.
- Interpreting module and cell measures of performance.

Sometimes assembly and machining operations are combined, expanding the range of skills required by individual team members. It is not proposed that everyone in the cell must be fully competent in all areas, because some people may prefer to specialize in particular operations. However, real long-term benefits from effective manufacturing systems are dependent upon people possessing the broad skills necessary to work in multidisciplinary teams.

Information needed to support the selection process should be obtained through several routes in order to obtain a fair assessment of abilities and determine overall training needs. This information should be sought from an immediate supervisor, a self-evaluation questionnaire and appropriate written assessment tests, culminating in a formal interview. Assessment tests must be tailored to the particular job competency and skill requirements; they compare *what people say they do and what they actually can do.* Standard tests have been devised and are available for evaluating a wide range of competencies and skills. These tests must be administered professionally, and include giving positive feedback to candidates, helping them to identify training requirements. In my experience, the use of formal testing is crucial to obtaining a realistic picture of people's ability. Several years ago following a factory redesign, the manufacturing system did not show the expected performance improvements. As part of the post-implementation investigation, we checked the process capability of the plant finding 50 per cent incapability. We also tested the basic skills of the workforce and found that 50 per cent could not read drawings adequately or follow written instructions. It took considerable effort to recondition equipment to achieve full process capability and train cell team members in basic skills. Once accomplished, within a year the business won national recognition as one of Britain's best six factories by *Management Today* (September 1988).

The selection process culminates in a formally structured face-to-face interview to elicit evidence of the candidate's behavioural competencies, technical skills and manufacturing knowledge, ultimately aimed at assessing his/her suitability for membership of a particular cell team. Questions to be asked at the interview should be agreed by the selection team prior to the event with sufficient time allowed for appropriate answers. Using standard questions allows interviews to be conducted by different managers, while still being able to compare people's abilities, selecting the best candidates for cell teams.

> *A rigorous selection process for prospective cell team members is critical; if someone does not have the basic aptitude or commitment to adopt new working practices, the chances of achieving a world class manufacturing system are remote.*

The interview and selection teams should hold a feedback session with each candidate following the appointment of team members, providing constructive comments on how personal performance could be improved. This should be linked to preparing individual personal development and training plans mapping how people will be trained enabling them to achieve their full potential. Those unsuccessful in securing a new role within a cell team must be treated with dignity and given a full explanation as to why they have not been selected. They should be given every assistance, sometimes in difficult circumstances, in finding alternative employment opportunities.

Continual improvement process

All conceptual designs can be improved by development and this includes manufacturing systems. The systematic design of manufacturing processes is fundamental to creating an effective supply chain, but world class manufacturing systems can only be achieved by introducing structured continual improvement programmes aimed at refining critical features needed to further enhance operational performance. The range of activities to be addressed normally exceeds available resources, but the management challenge is *how to encourage everyone within the organization to suggest and implement ideas for improving the process.*

Such an approach dramatically increases the available resources – a team's thinking power is always greater than that of the brightest individual. The module leader's task is to identify and promote projects having the greatest impact upon customer satisfaction, quality and operational performance. These must be supported by senior managers and resources made available, structuring the way people work together to improve operational performance. The continual improvement process has several stages, following traditional routines of identifying problems, devising plans, implementing solutions and checking that the solution works. (Fig. 6.1).

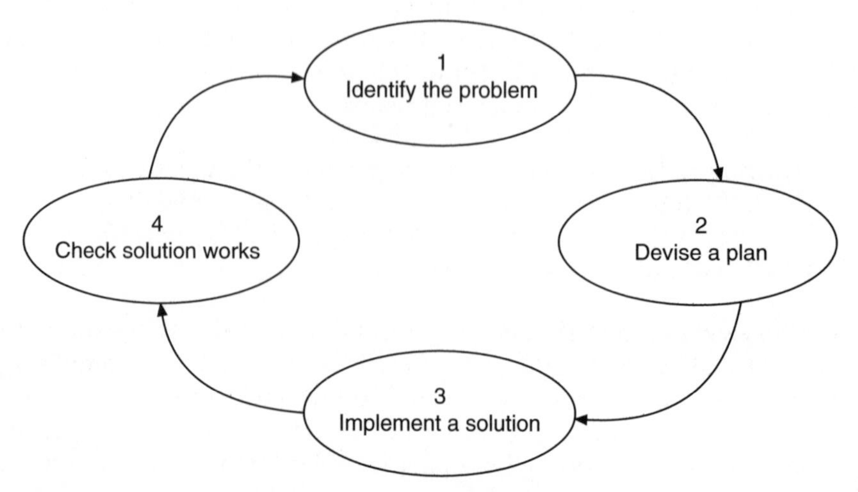

Figure 6.1 Structured continuous development process

The stages in the continuous improvement process (Table 6.1):

Stage 1 – Identify the problem

- Identify the area needing improvement.
- Confirm it will improve customer satisfaction, quality or operational costs.
- Define the actual problem, quantifying magnitude of identified changes.
- Collect information related to problem and determine relationship between cause and effect.
- Analyse the root cause and ask 'why?' five times to confirm findings.
- Develop possible solutions, assessing relative merits to determine most feasible alternative.
- Confirm possible benefits and if necessary complete risk analysis on preferred solution.

Stage 2 – Devise a plan

- Identify actions to be taken to implement proposal.
- Assess resources needed to complete the different tasks.
- Prepare a cost estimate for implementing changes.
- Determine the time scale to complete various tasks.
- Obtain any necessary approvals to implement proposals.
- Assign responsibility for the tasks to team members.
- Develop measures of performance to confirm problem has been resolved and corresponding performance improvements achieved.
- Prepare an agreed task and action list to record and monitor progress.
- Assess training requirements and ways of acquiring any newly identified skills.

Table 6.1 Continuous improvement action list

Continuous improvement – actions						
No.	*Date*	*Task*	*Proposed action*	*Person responsible*	*Date required*	*Date achieved*

Stage 3 – Implement a solution

- Confirm resources needed will be available to complete implementation.
- Agree an implementation plan, gaining module manager's commitment.
- Implement proposed changes.

Stage 4 – Check the solution works

- Monitor the changes and compare the results against previous situation, record identified measures of performance to demonstrate success.
- Critically assess and analyse results.
- Confirm all the expected benefits have been delivered.
- Determine possible reasons for any deviations to expected results.
- Identify further improvements that could be introduced to enhance performance.
- Implement corrective actions to formally record changes, if necessary.
- Revise standard operations sheet, and obtain formal approval for process changes, if required.
- Provide any training needed before adopting new process.
- Review revised process, record lessons learnt.
- Recognize and congratulate team on its achievements.

The most successful continuous improvement projects are those identified by the cell team when people organize themselves into groups to tackle specific problems leading to workable solutions. Effective continuous improvement projects also require management commitment in providing *time* for cell teams to think about issues, making resources available to implement proposed solutions.

Continuous improvement route maps

Identifying problems that need resolving following module implementation includes rectifying any obvious deficiencies in the supply-chain and manufacturing systems. These should be expected as part of the module/cell launch and the transition in working practices needed to establish good team working. However, once they have been resolved and operations settled into a consistent rhythm, formal continual improvement programmes must be introduced, as an established way of working. A more structured method for identifying areas requiring improvement can be developed using module and cell performance measures collected as part of the management reporting process.

The following parameters provide a basis for driving continual improvement programmes:

- Customer satisfaction.
- Adherence to customer delivery schedules.
- Product quality and number of engineering concessions.
- Cost of quality.
- Stock utilization and material flow.
- Team productivity.
- Direct costs controlled by module/cell team.
- Utilization and availability of bottleneck processes.

Actual cell performance can be monitored against the operating targets established during the manufacturing design phase, to determine any significant areas of underachievement. These must be critically assessed and priorities assigned to parameters that could be the focus of continual improvement groups. It is imperative that operational performance targets are achieved rapidly following implementation in order to support the financial justification and secure future commercial viability. One method of structuring the continual improvement programme is to develop a *route map*, identifying key factors that directly influence module/cell performance. The cell team and the module supervisors must work together to determine priorities for the continual improvement programme, but those items the team believe they can resolve on their own should always be given preference. Successful implementations demonstrate achievement and increase confidence in dealing with more complex problems.

Adherence to customer delivery schedules is a common problem following any factory reorganization and a typical route map is shown in Fig. 6.2.

Introducing successful continuous improvement teams requires considerable management involvement and intuition; teams identify problems and people are generally committed to resolving any problems which prevent the cell from achieving its full potential. Traditional production management methods did not encourage participation in solving problems. These same managers must now adopt an entirely different approach, ensuring that people understand the benefits of working together in teams, and providing the necessary leadership to make continuous improvement programmes successful. In some instances the selection process may have identified a manager's reluctance to embrace these new concepts. It is imperative he or she is not appointed into a module management role.

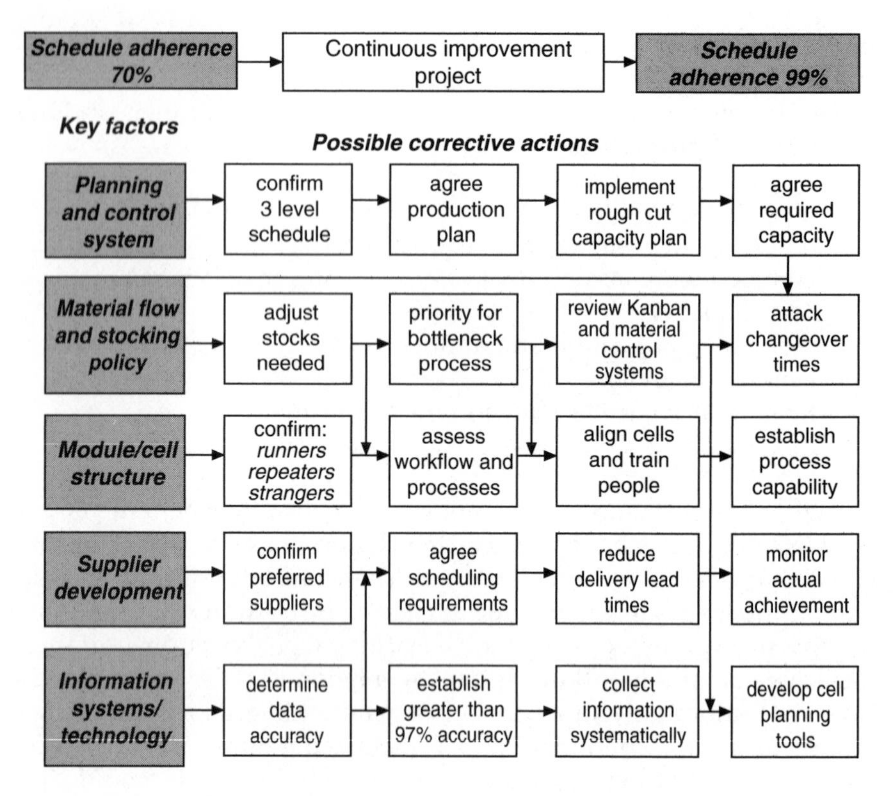

Figure 6.2 Continuous improvement route map

> *Managers must be totally committed to implementing modular manufacturing systems and demonstrate their willingness to accept changes in working practices, or they have no place in the company.*

Requirements for continuous improvement groups to work effectively include:

- Senior managers who demonstrate full support for the programme and promote achievements.
- Groups formed through common understanding, not exhortation.
- Managers who encourage groups to take the initiative, providing a trusted facilitator, if required.
- Cell teams must be trained in topics needed to support successful continual improvement projects.
- Normal working time must be allocated to continuous improvement projects.
- Meeting place should be available for group discussion and displaying project plans.

- Managers must never direct group deliberations, listening to recommendations.
- Budgets should be established to provide necessary funding.
- Resources made available to implement the group's suggestions.
- Technical expertise provided, when required.
- Group leader assigned responsibility for coordinating activities.
- Fewer than ten people in group.
- Times of group meetings should be agreed.
- Action plans should be prepared and circulated by team.
- Mutual trust must be created with common goals for delighting customers and delivering financial commitments.
- Achievements should be formally acknowledged.

The introduction of continuous improvement teams, following the successful implementation of a rigorously designed manufacturing process, is the only way for a business to become truly world class. In isolation, neither manufacturing design nor continual improvement initiatives can deliver the ultimate operating performance so everyone in the company must strive to make these two initiatives mutually supportive and the accepted way of creating world class manufacturing operations. When team-orientated manufacturing systems, based upon modular organizations and linked to continuous improvement groups, become the normal way of working, it is incredible what can be achieved. I have been involved with several businesses in Europe and the United States of America that successfully made the transition from traditional military contractors into high quality, commercial aircraft systems suppliers. They designed and implemented manufacturing systems for products that other companies found unprofitable, and through committed teamwork developed excellent operational processes which transformed them into world class aerospace system suppliers.

Continuous improvement training

To be successful, manufacturing teams and continuous improvement groups must be trained in effective manufacturing concepts and basic analytical skills, providing an understanding of the manufacturing principles and systems approach adopted when designing component manufacturing and assembly modules.

Managers cannot expect people to readily accept the introduction of modules or make meaningful contributions to continual improvement programmes without detailed explanations and training in these alternative ways of working.

A set of formal training sessions should be devised, tailored to the specific needs of the business, educating people to work in teams and assisting them to acquire a broader range of technical skills. This task of training the workforce is probably the most significant commitment a senior management team has to make. It will:

- Take people away from their place of work.
- Consume considerable 'normal' production time.
- Lower the efficiency of specialist operators while they train other team members.
- Increase the training budget to pay for additional support.
- Make people more employable by other companies.

However, without this concerted effort to train people in new skills, people may not be able to work effectively in teams, or deliver the operational performance improvements expected from adopting modular manufacturing systems.

The training programme should be structured to provide fundamental underpinning knowledge pertinent to launching modular manufacturing systems and techniques for securing improved operational performance. Additional training programmes can be scheduled at regular intervals (say over the next three years) to meet individual personal development and training plans. Topics that need to be covered depend upon several factors, for instance level of present knowledge, type of business, position in the supply chain, current manufacturing methods, magnitude of step change being introduced and the overall business objectives, therefore making a bespoke range of subjects for every organization. A set of core topics can be identified as a foundation for modular manufacturing system training.

Modular manufacturing

Understanding customers

All businesses must take care of its customers, because without them they cannot exist. The workforce must understand that it is considerably easier to maintain and satisfy existing customers than to develop new ones; meeting and exceeding customer expectation is of paramount importance.

The topics to be covered include:

- Measurement of customer satisfaction:
  - assesses how the company's products and services are perceived

to perform in service. Standards must be established based upon customer expectations and monitored regularly to confirm they are actually delivered.

- Service expected by the customer:
  - customers require identified interfaces allowing simple communications;
  - orders to be accepted with confirmation of delivery dates;
  - items to be delivered on time, quality assured and in the correct quantities;
  - problems to be resolved quickly, efficiently and in a friendly manner;
  - their own systems/procedures to be adopted and supported.
- Reasons for customer dissatisfaction:
  - manufacturing quality problems on parts received from suppliers;
  - problems or errors preventing installation of components;
  - late delivery;
  - mistakes in the product specification;
  - failure to meet product performance criteria;
  - component variability on assembly;
  - product unreliability in service;
  - supplier 'arrogance';
  - failing to listen to customer comments/requests.
- Customer requirements:
  - customers define quality standards and expect compliance;
  - company must provide the level of service expected by customer;
  - formal points of contact to quickly resolve issues;
  - proactive response to possible problems;
  - open communication and partnership.
- Managing customer expectations:
  - company image projected through advertising, promotions, technical publications and press articles;
  - customer demands must be effectively managed, ensuring they can be consistently met by the business;
  - promises and commitments must be capable of being delivered;
  - customer expectations should be quantified and used to monitor actual performance.
- Who is my customer?
  - customers are the next team in the supply chain;

- customers can be internal or external, but they must be identified;
- everyone needs to know their particular customer's requirements;
- customer requirements should be mutually agreed;
- teams should seek feedback from their customer;
- customer views should be documented and discussed;
- team members should take positive action to improve customer relationships.

- Fake customers:
  - create an interface without adding value;
  - promote routines because of custom and practice;
  - exhort and deliver rhetoric without details of actual requirements.

- Suppliers:
  - provide products and services;
  - need to understand specific requirements and priorities;
  - require honest feedback on actual performance;
  - are involved in establishing actual requirements;
  - help build mutual trust and partnerships;
  - need open communication and simple systems.

- Integrated value chains:
  - everyone has a role as a supplier and a customer;
  - activities linked in a series of customer/supplier relationships;
  - chain is only as strong as the weakest link;
  - everyone must understand each other's requirements, capabilities and limitations;
  - teams must work together to optimize the total process, not individual elements;
  - relationships should be managed to provide mutual benefit;
  - everyone must strive to make relationships and partnerships work effectively.

Manufacturing processes and systems
The manufacturing system design and plan for transforming the business into a process driven, team-based organization must be condensed into a concise, understandable training package, based on the following information:

- What is a process?
  - series of activities which converts an input into an output;

- introduction to core business processes;
- need for specific business processes;
- requirements at different organizational levels;
- core processes identified for the business to retain on site;
- outline of make versus buy decision;
- details of key manufacturing processes to be performed on site;
- details of proposed module and cellular structures.

• Attributes of good processes:

- capable of meeting customer requirements;
- flexibility to respond to change;
- robust to consistently achieve objectives;
- effective, meeting overall cost targets;
- efficient, making optimum use of resources;
- product and process control attributes identified;
- capable of operating within required tolerance limits;
- capability monitored using statistical methods;
- measuring systems known to be capable.

• Module and cell requirements:

- people selection and team working;
- processes and standard working practices defined;
- cell and team structures designed to perform particular activities;
- smooth materials flow between cell processes and modules;
- Kanban and material control systems;
- importance of lead time, differentiating between processing/ waiting time;
- support groups needed for key processes and teams;
- continuous improvement groups contribute to meeting stakeholder expectations.

• Role of team members:

- multi-skilled and capable of performing a variety of tasks;
- commited to continual development;
- responsible for own quality;
- participate in cross-training other team members;
- progressively reduce lead times;
- continually attack set-up times;
- commited to the elimination of waste;
- active participation in continual improvements.

- Mapping and understanding processes:
  - identify process boundaries;
  - define actual process;
  - determine inputs and outputs for process;
  - analyse current methods to determine non-value added activities, bottlenecks and other limitations on performance;
  - identify improvements to processes;
  - understand time-based process mapping techniques;
  - determine the most effective way of operating the processes;
  - document the recommendation and quantified benefits;
  - prepare standard operating procedures.
- Key factors and measures impacting cell performance:
  - delivery performance against customer schedules;
  - number of defects identified in cell and by customer;
  - cost of quality – scrap and rectification, quality assurance and warranty;
  - customer perception, including both internal and external customers;
  - supplier performance – quality, on-time delivery and cost;
  - stock utilization in cell, module and supply chain;
  - production volumes and changes in product variety;
  - controllable costs that can be influenced by cell teams.

Operational improvement

Elimination of waste

All operations contain elements of useful value added work and also waste. Japanese car manufacturers developed ways of identifying specific types of waste eliminating them through continual improvement programmes (see p. 91). Techniques to remove these common items of waste must be understood.

- Defective products and mistakes.
  These are expensive in time and materials. The aim should be to produce 'right first time' through checking materials, tools and operating procedures prior to commencing the job. Attacking defects improves:
  - productivity, making more products with the same effort;
  - quality, reducing rework;
  - effectiveness, using less resources for each operation;
  - morale and pride in products.
- Overproduction.

Make what is required when it is needed. Overproduction:

- increases work in progress, tying money into stock;
- hides problems embedded in process;
- creates inventory before customer is prepared to pay for it;
- consumes resources that may be required for a more urgent job.

- Operational processes.
  Not all operational processes add value to the product or service; ones that do not add value or are required by the customer must be eliminated. These:
  - increase number of stages to complete jobs;
  - add cost to product;
  - consume unnecessary resources;
  - increase documentation and bureaucracy.
- Transportation.
  This never increases the value of the product, it merely:
  - increases non-productive cycle times;
  - prevents a smooth flow of work;
  - creates possible bottlenecks;
  - increases costs of equipment and factory space;
  - causes damage to components.
- Unnecessary movement.
  Wastes time and energy without adding value, manifest by:
  - people walking too far between operations thus wasting time;
  - workplaces not ergonomically designed, adding to cycle times.
- Excess inventory.
  Inventory costs money and should be regarded as 'evil', not an asset because it:
  - ties up capital that could be more profitably deployed;
  - requires space and appropriate environment for storage;
  - requires information system to record and locate stocks;
  - suffers losses, devaluation, deterioration and obsolescence;
  - incurs people's time to protect, supervise and administer.
- Waiting for resources.
  Waiting for materials, equipment or people increases lead times and costs, as it:
  - decreases productivity;
  - interrupts the flow of work;
  - impacts total plant capacity on bottleneck processes;
  - creates disruption to related activities.

Cleanliness and good housekeeping

Good housekeeping is fundamental to creating a conducive working environment and making people take pride in their workplace. The Japanese introduced the *five Ss* as an effective way of improving housekeeping and promoting good working practices. The five Ss are Seiri, Seiton, Seiso, Seiketsu and Shitsuke. No direct translation exists, but the concepts are broadly:

- Sort – throw out everything that is not directly necessary for current operations, including equipment, documentation, components, tools and such.
- Segregate – organize and arrange items close to where they are used and establish finite limits on quantity required.
- Shine – clean the area, equipment and tooling, encouraging people to take care of facilities and eliminate possible quality and personal health problems resulting from dirt, dust and a poor working environment.
- Strengthen – adopt the first three items as a normal way of working, reinforced by encouraging problems to appear and be resolved by teamwork.
- Standard – insist on basic disciplines needed to ensure people follow procedures and work to standard operations.

Improving product quality

Identifying the root cause of problems

Another Japanese technique for identifying the root cause of problems is *five Whys*.

This insists on people asking and answering the question *why?* five times for every problem.

By the fifth time the underlying issues should have been identified:

1. Why ?
2. Why ?
3. Why ?
4. Why ?
5. Why ? Yes!

Controlling processes

Statistical process control (SPC) is the established technique for controlling processes within required tolerance bands. The challenge is to prevent and reduce variation, making the process predictable. This is achieved by measuring the key characteristics of the product throughout the manufacturing process to determine the degree of variation and the extent to which it falls within the limits specified by the customer, as opposed to inspecting the product when finished. SPC techniques are well documented and based upon two types of variables known as 'average' and 'range'. These variables are plotted onto charts, tracking the average measurement and the range of maximum and minimum values for those parameters identified as key characteristics. Samples of between five and 20 components are taken at predetermined intervals and used to calculate the average and range parameters. They are plotted on a chart showing the tolerance band with the upper and lower control limits that have been calculated for a particular characteristic, based upon the sample size and statistical constants required for a desired level of process capability.

Operators examine the patterns created by the measurements and interpret possible circumstances leading to specific plot formations. These plots can be used to prevent parameters from moving outside the control limits. The objective is to ensure all parameters remain within the upper and lower control limits, confirming process capability. This is calculated using a standard formula and is expressed as a Cpk; it should be greater than 1.33 if the process is capable. Cpk measures the capability relative to the specification limits, as opposed to the process capability measure Cp that compares tolerance spread with the process spread. Under normal operating conditions Cpk is the more important parameter, because it is possible to have a well-controlled process outside the necessary tolerance band. With experience, the manufacturing teams should be able to use the information recorded on these charts to identify ways of reducing variability, striving to restrict the range of measurements and move average readings towards the mean of tolerance band.

A simpler method for controlling manufacturing processes is called 'Precontrol', devised by Peter Shainin. This involves constructing a time-based chart that monitors variation and provides visual assistance for maintaining key parameters at a safe point within the tolerance band. The rules of Precontrol are relatively simple:

- Determine the Precontrol limits by taking the total tolerance band and dividing it into four equal parts. The centre two sections around the mean tolerance are green, the two outer sections yellow, and outside the tolerance band is red (Fig. 6.3).
- The process is set to run at mean tolerance and parts measured until five consecutive pieces fall within the Precontrol limits (green bands).
- The process is then started measuring two parts on a recurring basis:
  - if both green → continue;

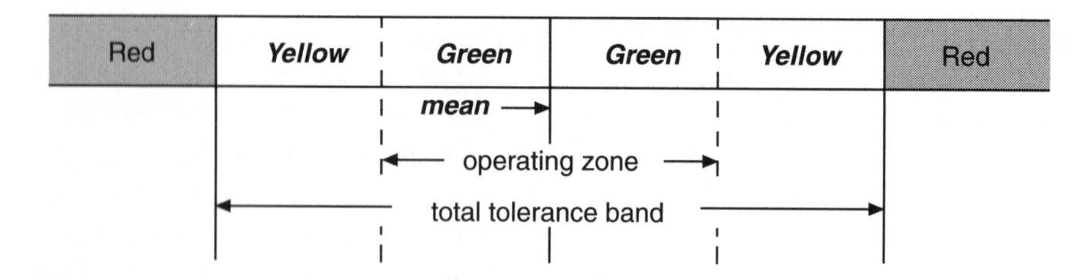

Figure 6.3 Precontrol parameters

  - if one green and one yellow → continue;
  - if both yellow → stop and identify the cause;
  - if one red → stop and investigate.

- Determine the time between stoppages to make adjustments, and set the sampling time interval at one-sixth of the time between subsequent adjustments; for example, sample every ten minutes if the process needed adjusting after one hour.

This method is often referred to as 'traffic light charts'. It is not as precise as SPC but the technique can be implemented quickly and provides an introduction to process control and machine capability. In many instances Precontrol provides sufficient information to ensure that processes are capable, making it unnecessary to introduce SPC.

The capability of measuring systems has a significant effect on the overall capability of a process; if they are incapable, available processing tolerances are eroded. Measuring systems should be an 'order of magnitude' more precise than the parameter tolerances being measured. If this cannot be readily achieved then the measuring technique must be changed.

The suitability of measuring systems can be determined using:

- Gauge Repeatability and Reproducibility (Gauge R&R) studies to confirm the accuracy and repeatability of the measuring system for the parameters being measured.
- Isoplots using two measurements taken from several components. The capability can be confirmed by plotting these parameters and measuring the size of the lozenge, Peter Shainin technique (Bhote 1988).

I have not attempted here to provide the detailed information needed to train people in these statistical techniques, merely to raise awareness of how important it is to control processes. These techniques have only limited use in identifying reasons for variation or for establishing corrective actions, but in my opinion they were the most significant factor in transforming the quality and competitiveness of the motor component industry. (When your main customer informs you that 50 per cent of your processes are incapable and gives you two months to achieve better than 95 per cent capability or they will re-source the business, the rate of change and overall improvement in manufacturing practices that you can achieve is remarkable!)

Lead time reduction and material flow

The flow of materials through the factory is crucial to modular manufacturing. The aim is to establish a consistent flow of work, matching customer requirements. Topics that need to be covered include:

Level scheduling
This technique is used to smooth out workload variations. The production plan is rearranged to remove large batches, exploiting increased flexibility to manufacture smaller batch quantities more closely, aligned to actual daily or weekly customer requirements. The production volumes and associated changes in product mix are smoothed over short time intervals to fully utilize production capacity, promoting the manufacture of small, mixed batches of components. The objective is to establish regular repeat patterns of work that meet customer demand on a daily or weekly basis.

Changeover time reduction
Short changeover times are the key to low volume manufacturing and crucial for supporting mixed mode, make daily/sell daily production. Managers stressing the importance of short changeover times, and instigating a team effort on developing alternative methods, can achieve considerable changeover time reductions. The technique for reducing changeover times is to identify internal operations that can only be performed when the machine is stopped and external operations that can be carried out while the machine is running. Performing external tasks prior to stopping the machine obviously saves time. Closely observing current methods using a video camera also gives a useful insight into how time is being wasted performing non-productive activities during equipment changeover. These make good continuous improvement projects because achievements can be monitored demonstrating progress. Regular practice using simple tools usually delivers dramatic results.

Kanban systems
Kanbans provide a link between the benefits of small batch sizes, quick

changeovers and visual controls needed for reducing lead times. A Kanban is a signal that passes customer demands down through the process (p. 94). It also limits the numbers produced to match customer requirements. Kanbans are used to promote the flow of work between workstations where physical distance prevents a consistent flow of materials, acting as a conveyor linking material supplies, parts manufacture, assembly and test to the customer. Simple rules have to be established for operating a Kanban:

- The assembly process releases Kanbans and pulls the correct number of parts from manufacturing cells or suppliers based on Kanban card information.
- Suppliers receiving Kanban provide parts according to Kanban information.
- No work is started until the Kanban signal is received.
- Kanban cards accompany containers used to protect and transport components.
- Everyone is responsible for supplying the correct number of quality assured parts and any defects are rectified prior to delivery.
- The number of Kanbans in circulation for each component is gradually reduced to expose problems and remove any excess stock in the supply chain.
- Problems made visible through a reduction of work in progress must be resolved through identifying the root cause and implementing corrective actions.

These areas identified for initial training do not cover all the critical learning aspects required by the workforce, nor is the training complete when such topics have been formally taught. The financial and time commitments needed for training people must be fully accepted by the management team, while appreciating that:

> *A skilled, well-trained, flexible workforce will always outperform one which uses traditional methods.*
>
> *Training is an ongoing investment; many high performing benchmark companies have taken several years to develop the versatile workforce necessary to compete in world markets.*

Design of experiments for problem solving

Improving the robustness of supply-chain and manufacturing processes is dependent upon identifying and solving problems. Statistical process control, FMEA, cell audits and continuous improvement groups all identify problems. In many instances the solution is relatively straightforward and can be determined using five Whys or root cause analysis. However, in particular circumstances when several factors interact, identifying the solution becomes more difficult and requires alternative techniques to resolve problems. These methods are based on the design of experiments that systematically isolate the variables causing the problem. The best known of these techniques is the Taguchi method that minimizes the number of changes required to each variable when searching for the optimum combination of parameters. Factorial tables are used to reduce the number of experiments and find the performance plateau that avoids high performance sensitivity to small changes in particular variables. Taguchi methods are well documented, but they do require specialist knowledge and should be applied after other techniques have failed to provide a solution.

Peter Shainin has devised a more usable design of experimental techniques. These are aimed at finding the cause of the variation that Shainin calls 'Red X'. The steps to finding the Red X are:

- Find the root cause of any variation (possible Red Xs) having a detrimental effect upon the product.
- Define the correct tolerance range for each Red X to meet customer expectations.
- Control each critical key characteristic (Red Xs) to maintain these parameters within the specified tolerance bands.

Identifying Red Xs is complicated by possible interaction between different variables; in some instances the second or third largest causes are sufficiently large to be of practical significance, referred to as the 'Pink X' and the 'Pale Pink X'. Shainin has developed a family of analytical techniques: reducing the number of variables to be considered, identifying suspect variables into a small group of factors before completing a designed, full factorial experiment:

Multi-vari charts

- Seek to find variations within a part:
  - end to end;
  - right, left and centre;
  - specific features; and
  - time period to time period.

A number of samples are taken in order of production with variations grouped into families in order to categorize different variation types. These patterns are analysed and sorted into those factors having greatest impact upon the available tolerance range. Actions are then taken to systematically remove the different forms of variation from the process.

Paired comparisons

- Uses detailed measurements from different parts to determine which features have the greatest impact upon the performance of the unit:
  - inspect five sets of two or three consecutively manufactured parts – record the findings as good, bad, good or one good and one bad;
  - measure and record as many parameters as possible;
  - compare values within the pairs looking for consistent changes in particular values.

If components are used in a product assembly that can be tested, the impact of interchanging a good part with a defective one can be assessed until the actual 'bad' component that causes the unit to fail is identified.

Scatter plots

- Examines the changes in product characteristics for a particular product variable:
  - select 30 random examples of units made to the current tolerance;
  - measure the product or process variable and record the effect upon the product characteristic;
  - plot the product characteristic against the product variable;
  - examine the shape of the pattern made by the plot.

The overall shape, grouping and position of the points show the relationship between the product characteristic and the product variable. For example, if the points form a line then a linear relationship exists between the two factors.

Precontrol (see p. 251)

- Monitors the variation of key product characteristics throughout the manufacturing process with respect to time:

- measure two components at a cycle time interval of one-sixth of the time between subsequent process adjustments;
- plot readings with respect to time on the control chart;
- review the shape of the plots.

The movement in parameters being measured shows trends, identifying possible causes of variation.

Full factorial experiments

- Searches for the cause and effect relationships between different parameters:
  - identify matrix of parameters that can be varied and are thought to have an impact upon the product;
  - select one variable and measure the effect of changing the parameter, repeated for (say) three different values;
  - take another variable, measure the effect and repeat for different values;
  - continue with the matrix of parameters until each variable has been completed.

Plotting the effect of changing each product variable provides the degree of scatter and the significance each parameter has on performance. It must be remembered that full factorial experiments for over four variables involve a considerable number of experiments and should only be used once the critical factors have been identified using other methods.

Detailed information on how to apply these techniques can be found in *World Class Quality* (Bhote 1988), or from Shainin Consultants, Inc., Connecticut, USA.

Module audits

These should be introduced by the supply-chain manager to demonstrate commitment to providing the process capable facilities needed by manufacturing teams to produce high quality products on time and to ensure that people's efforts concentrate upon those factors important to the business. The supply-chain manager, accompanied by the module manager, should make formal visits to each cell, auditing its status against an agreed checklist. Visits should take place at random intervals and at short notice.

The seven items to be audited include:

1. Quality performance information available in the cell:

   - scrap details by part number, type of defect and frequency of occurrence;
   - rectification information by part number, corrective actions taken and frequency of occurrence;
   - use and display of check sheets to ensure compliance to process;
   - validity of quality information, process details and product documentation.

2. Equipment maintenance:

   - display and recording of preventive maintenance routines;
   - status of process capability checks and actions taken to reduce variability;
   - warning and safety notices on equipment;
   - removal and handling of waste including any byproducts of the process;
   - state of service and auxiliary items needed to operate equipment;
   - oil leaks or other contaminants seeping onto the floor;
   - overall condition of machines – clean, no loose bolts or obvious faults.

3. Safety:

   - safety guards in good working order and being correctly used;
   - dust and fume extraction fitted, if required, and in good working order;
   - noise levels acceptable, excessive sources of noise screened and attenuated effectively;
   - good lighting levels, exceeding legal factory requirements;
   - floors are clean and free from liquid deposits oozing from equipment;
   - fire extinguishers are available and positioned accessibly;
   - health, safety and environmental notices prominently displayed with team members heeding instructions;
   - non-obstruction of entrances, exits and emergency exits;
   - non-obstruction of gangways and routes used for transporting material;
   - adequate provision of first aid equipment and trained personnel;
   - people use necessary protective equipment – safety glasses, safety shoes, ear protectors and such;
   - adequate provision of safety equipment for emergency use.

4. Housekeeping:
   - aisles and gangways clear of work in progress, packaging and other obstacles;
   - materials handling equipment suitable for purpose and safe;
   - workplace is clean and appropriate to tasks being carried out;
   - no materials, tools, equipment or waste placed on floor;
   - floor adequately painted or finished to allow easy cleaning;
   - equipment painted and in good overall condition.
5. Container control:
   - containers used to store and transport components are clean and protect products from the environment;
   - Kanban storage locations and transport loops are in operation;
   - Kanban cards and other 'shop traveller' information in correct place and adequately protected;
   - quantity of components correct and in designated container;
   - components in the container are of acceptable quality;
   - any items that for any reason cannot be placed in a container are properly identified and corrective action instructions documented;
   - containers with kits of parts have the correct mix and number of components.
6. Continuous improvement:
   - status of continuous improvement initiatives;
   - actions planned, documented and implementation agreed;
   - reduction in changeover times on equipment monitored;
   - savings on quality costs displayed;
   - progress on lead time reductions documented;
   - cost reductions through the elimination of waste identified.
7. General:
   - time spent training team members recorded;
   - records of training needs and individual achievements displayed;
   - key topics that remain outstanding for team training identified with dates for training established;
   - notice boards are used, neatly displaying current information;
   - clean uniforms worn by cell teams;
   - orderliness of cell of a high standard;
   - state of work benches and equipment – clean and equipped with labour-saving devices;
   - storage system for information and documentation is accessible, maintaining it in good condition.

The items to be audited need to be discussed and agreed between the auditor and the cell teams to set the standards against which parameters will be judged. The standards should be challenging, but if set too high they may never be achieved and have a demoralizing effect; if set too low they will be readily achieved, giving no sense of achievement or motivation to improve. *The standards set must be seen as stretching but achievable with effort.*

The audit operates by scoring the seven headings, awarding points on a scale, say, of *1–10*:

1–2 Unacceptable – needs immediate action.
3–4 Below average – plans for improvement required.
5 Average – must agree corrective actions.
6–7 Acceptable – will strive to improve.
8–9 Above average.
10 Good.

Items that score particularly high or low marks must be quantified and remarks included in a summary report. The marks for each of the seven headings are averaged to provide an audit score for the cell team, and displayed on the cell notice board in graph form, showing improvements made from previous audits. Features from audits identified as requiring improvement should also be displayed with action plans and implementation dates. It must be remembered that cell audits are a two-way process; corrective actions may require operational management support plus additional expenditure to rectify deficiencies. It is imperative that operational managers demonstrate their full commitment to the audit and continual improvement process by ensuring manufacturing support teams respond to the cell team's requests in a positive and timely manner. Corrective actions outstanding at the time of the next audit must be seriously questioned and reasons established why people failed to carry them out.

Module information

It is important to provide cell teams with accurate, timely information that increases awareness about the state of the company, customer performance, main change projects and how the module/cell is performing against its targets. It should be updated weekly and displayed prominently as a balanced score card (Fig. 6.4), creating a focal point where team members discuss issues with the module leader and review key factors important to the business and module.

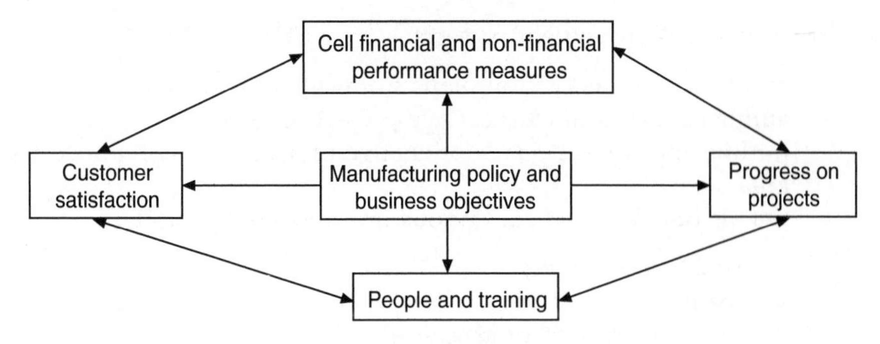

Figure 6.4 Balanced score card for manufacturing cells

The information must be tailored specifically to meet module and cell team requirements and include statements outlining:

- Summary of the manufacturing policy and overall business objectives:
  - critical issues impacting customers;
  - present market situation and expected trends;
  - relevant issues identified in business plan;
  - projected market share for main product lines;
  - competitors/competition in the particular market segments; and
  - performance targets to meet business plan commitments.
- Customer satisfaction and delivery performance achievements:
  - customer measures based on quality and delivery;
  - delivery to schedule by product line;
  - scrap and rejects by part number; and
  - cost of quality for cell.
- Cell financial and non-financial performance:
  - sales value of items produced in cell;
  - value of stock and work in progress;
  - added value per employee;
  - added value per unit of pay;
  - manufacturing costs controlled by the cell:
    - ◊ overtime payments;
    - ◊ consumable materials;
    - ◊ production materials;
  - percentage of equipment known to be process capable;
  - average lead time in cell;
  - percentage of unplanned downtime/available hours; and
  - other specific measures crucial for running the cell.

- Progress on improvement projects:
  - marks scored and action plans from cell audits;
  - summary of significant change projects on site;
  - outline and project plans for change projects that will affect the area;
  - details of cell-based continuous improvement projects:
    - ◊ project objectives;
    - ◊ team members;
    - ◊ action plan with time scales;
    - ◊ named people responsible for actions;
    - ◊ progress and achievements to date;
    - ◊ outstanding issues that need to be resolved:
  - investments in new plant and equipment; and
  - progress on significant new product introduction programmes.
- People and training:
  - module and cell team members;
  - training requirements and skills matrix;
  - impact of time lost through absenteeism;
  - recognition for employees who have performed beyond expectation;
  - job opportunities to work in other areas and expand range of skills; and
  - social activities and events.

Information must be relevant and presented in a 'user-friendly' form. Graphs showing improvement should exhibit upward movement and comments written in understandable language. The more information that is provided, the more people relate to the business objectives, feeling part of the overall success and gaining increased job satisfaction. The balanced score card should show the supply-chain and manufacturing systems design are providing the long-term benefits envisaged at the project outset. The results following the introduction of the new system should quantify the step change in performance that has been achieved once systems are stable and module teams adequately trained. The introduction of continuous improvement initiatives must now be used to refine this detailed system design and overall processes developed until they are finally acknowledged by the industry as having benchmark performance that other companies would wish to emulate.

Conclusion

I have intended, and hope succeeded, in providing practical information for managers and multidisciplinary project teams on the sequence of logical events needed to improve operational performance through engineering effective manufacturing solutions. The book brings together the design and implementation of modular supply-chain and manufacturing systems, describing the various analysis phases that must be completed to design the optimum solution. I believe the design of effective internal manufacturing processes is critical for establishing a company's profitability and its ability to compete in global markets. Manufacturing is the real 'wealth creator' as it transforms market development and product introduction efforts into actual components, consistently 'adding value' to materials, converting them into products customers 'pay for'. Manufacturing resources account for the greatest part of management and financial assets needed to run a business, but traditionally little effort has been directed towards creating good factories. Many senior managers have not appreciated the wide range of tasks involved or recognized the enormous benefits that can be gained from implementing effective operating systems. External consultants have also been guilty of underrating the amount of work involved in designing world class factories, promoting an impression that this work can be completed very rapidly at shop floor level.

I have identified the supply-chain and manufacturing system design techniques used in successful implementation projects, presenting them under the following chapter headings:

Identify the need for change – translating the company business plan into operational targets that must be delivered through designing core processes, confirming financial commitments are achievable.

Module/cell identification – developing a manufacturing strategy, identifying gaps in current operational performance and by applying strategic make versus buy analysis, identifying the manufacturing activities and associated modular structures needed to be a successful business.

Steady state design – determining the size of modules/cells needed to satisfy market demand and designing how they must operate under steady state conditions, taking into account quality requirements, benefits of applying Japanese techniques, introducing new technology, job definitions and module support services.

Dynamic design – making modifications to the modular structures orientating them to accommodate variable factors affecting operational performance, also selecting appropriate supply-chain, manufacturing planning and control systems.

Financial justification – creating a supplier integration policy and developing a financial case so senior management will agree and support the implementation programme.

Implementation/continual improvement – installing integrated manufacturing modules and commissioning equipment. Defining the job requirements, selecting team members and providing essential training. Introducing continual improvement teams to enhance performance, monitoring achievements through meaningful measures of performance.

Teams follow a traditional route for introducing change:

- *Identify the ideal* – establish business targets, understand the current position, collect data, map the process and analyse the results.
- *Develop a concept* – design the process, test validity under a variety of operating conditions and make a proposal.
- *Define an implementation plan* – identify the resources and time scales, establish the costs and benefits and obtain management approval to fund implementation.
- *Implement the proposal* – identify job requirements and select cell teams, install facilities and equipment and train in new skills.
- *Check the results* – conduct audits and confirm expected benefits.

The requirements to accomplish each design stage have been addressed, which means similar themes are covered in different phases of the change process. The treatment of these topics is quite distinct as the project progresses from definition, to design, through training and implementation.

The job of designing effective manufacturing systems requires considerable resources from a dedicated team of knowledgeable people. In my experience, companies do not allocate sufficient resources to this most important task and risk obtaining partial solutions that are inadequately conceived, placing future company viability in jeopardy. A general manager must be able to identify the person responsible for designing how the factory will operate and having read this book, will understand more fully the challenge that is presented in designing supply-chain and manufacturing systems. In some instances the manufacturing policy or information needed to make informed judgements allows particular stages to be completed relatively quickly. I have provided diagrams with detailed checklists for an experienced manager to use and make informed assessments on features crucial to achieving superior operational performance. The benefit of systematically quantifying and qualifying fundamental operational requirements against the business plan objectives must not be compromised by glibly accepting traditional, possibly misinformed, opinions. Companies that operate to world class manufacturing standards consistently redesign their manufacturing processes to meet business needs and/or market requirements: this is the foundation of wealth creation. Once implemented, the supply-chain and manufacturing systems are refined using continual improvement programmes aimed at removing all forms of waste. It is the combination of system design linked to continual improvement that delivers world class performance; neither can deliver the benefits alone.

The transition of a traditional factory into an effective modular operation not only creates many opportunities for improvement, but also uncertainties and risk. Managers must be skilful at supervising the process keeping people informed and, whenever possible, involved in the design process. They must also give assurances on the advantages offered by these new ways of working, expressing realism about the consequences of failure. I have always found people are willing to accept the challenge of changed working practices constructively, provided that the business allocates sufficient time and resources to train them in the skills needed for their broader roles. This training commitment should not be underestimated; it bestows people with confidence to discard traditional practices and become effective team players. A decision to maintain the 'status quo' is not an option. In my experience, businesses that seriously accept the challenge of 'change' deliver incredible results, apparent in all areas of the business. People who work together in teams develop new levels of motivation and radiate the confidence to win both for the company and themselves. This ultimately provides lower product-manufacturing costs, excellent quality, delighted customers and the ability to compete with the world's best.

References and supporting literature

Bhote, K. R. (1988) *World Class Quality* (Cambridge, Mass.: American Management Association, Productivity Press).

Bicheno, J. (1998) *The Lean Toolbox* (Buckingham, England: PICSIE Books).

Bicheno, J. (1998) *The Quality 60, A Guide for Service and Manufacturing* (Buckingham, England: PICSIE Books).

Clarke, L. (1994) *The Essence of Change* (Englewood Cliffs, N.J.: Prentice Hall).

Drew, S. (1994) 'BPR in Financial Services: Factors for Success', *Long Range Planning*, Vol. 27, No. 5.

Ford Automotive Safety and Engineering Standards (1995) *Potential Failure Mode and Effects Analysis Handbook* (Dearborn, Michigan: Ford Motor Company).

Garside, J. (1998) *Plan to Win* (Basingstoke, England: Macmillan Business).

Hammer, M. and Champy, J. (1993) *Reengineering the Corporation* (London: Nicholas Brealey).

Hill, T. (1985) *Manufacturing Strategy* (London: Macmillan).

Japan Management Association (1989) *Kanban and Just in Time at Toyota; Management Begins at the Workplace*, trans. by D. Lu (Cambridge Mass.: Productivity Press).

Kaplan, R. S. and Norton, D. P. (1996) 'Using the Balanced Scorecard as a Strategic Management System', *Harvard Business Review*, Jan–Feb, Vol. 74, No. 1, p. 75.

Kaplan, R. S. and Norton, D. P. (1997) 'Why does a Business Need a Balanced Score Card?' *Journal of Cost Management*, May–June, Vol. 11, No. 3, p. 5.

Kennedy, C. (1994) 'Reengineering the Human Costs and Benefits', *Long Range Planning*, Vol. 27, No. 5.

Computer Sciences Corporation (1996) *The Manufacturing Industry Handbook*, CSC, Dog Kennel Lane, Solihull, England.

Parnaby, J. (1985) 'A Systems Approach to the Implementation of JIT Methodologies in Lucas Industries', *Int. Journal of Productivity Research*, Vol. 26, No. 3.

Parnaby, J. (1991) 'Designing Effective Organisations', *International Journal of Technology Management*, Vol. 6.

Parnaby, J. (1994) 'Business Process Systems Engineering', *International Journal of Technology Management*, Vol. 9.

Perera, T., Tipple, R., Mosley, J. (1996) *Effective Tool Management* (London: Crown Communications).

Perigord, M. (1990) *Achieving Total Quality Management* (Cambridge Mass.: Productivity Press).

Probert, D. R. (1995) *Make or Buy, your route to improved manufacturing performance* (London: Department of Trade and Industry, Produced by the Institution of Electrical Engineers).
Schonberger, R. J. (1982) *Japanese Manufacturing Techniques* (Cambridge Mass.: The Free Press).
Slack, N., Chambers, S., Harland, C., Harrison, A., Johnson, R. (1995) *Operations Management* (London: Pitman Publishing).
Time Management International (1992) *Putting People First*, TMI, Henley in Arden, Warwickshire, England.
Turner, F. (1993) 'Business Systems Engineering', *Proceedings of the Institution of Mechanical Engineers,* Vol. 208.
Walsh, C. (1993) *Key Management Ratios* (London: *Financial Times* and Pitman Publishing).
Wild, R. (1972) *Mass Production Management* (London: Wiley).
Womack, P. J., Jones, D. T. and Roos, D. (1990) *The Machine that Changed the World* (London: Maxwell Macmillan International).
Womack, P. J. and Jones, D. T. (1994) 'From Lean Production to Lean Enterprise', *Harvard Business Review,* March–April.
Yasuhiro, M. (1998) *Toyota Production System, An Integrated Approach to Just in Time, third edition* (London: Chapman and Hall).
Young, T. L. (1994) *Leading Projects, a Manager's Pocket Guide* (London: The Industrial Society).

Index